PRAISE FOR *INFECTIOUS*

'This book catapults us to the frontier of the vital science of infections and immune responses. Tregoning is a perfect guide, writing with wit and intelligence about a subject that surely everyone feels the importance of now. Brilliant and right on the zeitgeist.'

Daniel M. Davis, author of *The Beautiful Cure*

'What a book! A book for everyone, an expert, or an interested lay person, young or retired or somewhere in the middle. Informed, engaging, generous and superbly written. If you have secretly wondered about some of the information of the last year (as we all have) and want one book to explain it all with total clarity – this is the book for you. I started it and could not put it down. The best, most accessible, high-quality science book I have read this year. Utterly brilliant.'

Jeremy Farrar, director of the Wellcome Trust

'Tregoning is a gifted writer of popular science. He has a knack for explaining the intricacies of vaccines and immunity without dumbing them down, and he moves things briskly along with a barrage of often self-deprecating humour.'

Wall Street Journal

‘Packed with fascinating facts, intriguing anecdotes and more than a few Dad jokes, *Infectious* is an expertly guided, pacey tour through the world of all the stuff that’s trying to kill us and how our immune systems and human ingenuity are fighting back.’

Dr Kat Arney, science communicator
and author of *Rebel Cell*

‘To call this book timely would be yet another contender for understatement of this strange decade. As the pandemic has upended the world and ravaged the population, it’s a duty for all of us civilians to turn to experts like Dr Tregoning with due humility, and educate ourselves about what’s happening to us, how we got here, how we’ve dealt with similar events in the past, and how we might get through this. This book is thorough, engaging, entertaining and utterly vital.’

Frank Turner, singer-songwriter

INFECTIOUS

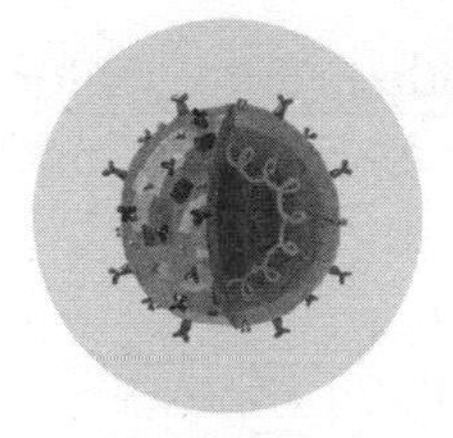
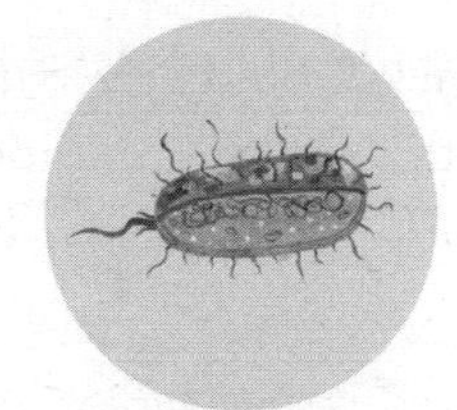
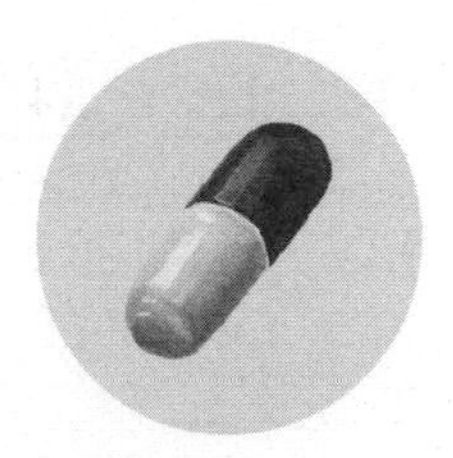

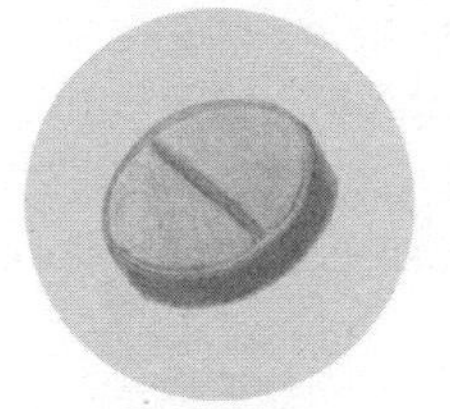
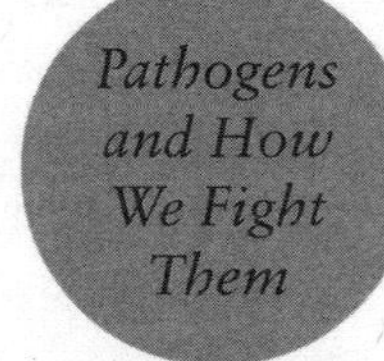

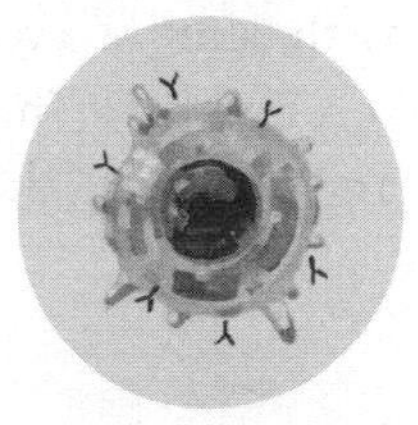

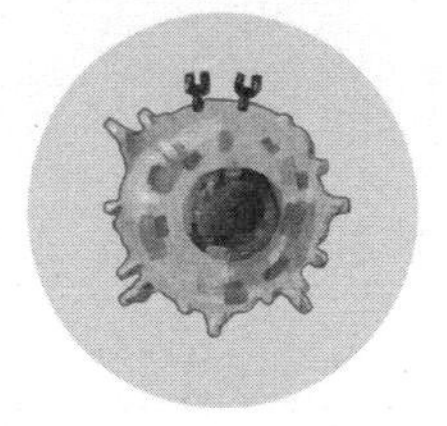

DR JOHN S. TREGONING

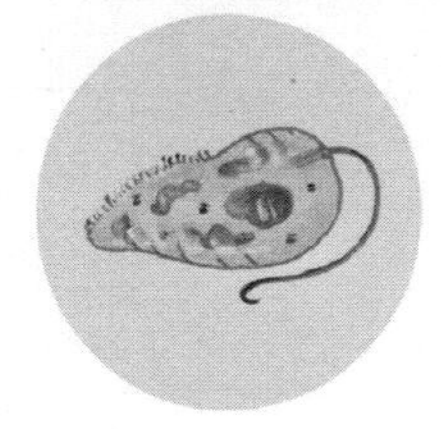

A Oneworld Book

First published by Oneworld Publications in 2021
This updated paperback edition published 2022

A CIP record for this title is available from the British Library

ISBN 978-0-86154-439-4
eISBN 978-0-86154-123-2

Typeset by Hewer Text UK Ltd, Edinburgh
Printed and bound in Great Britain by Clays Ltd, Elcograf S.p.A.

Some of this book has previously appeared on the Imperial College Medicine Blog, in *Nature*, and online for *Times Higher Education*.

Oneworld Publications
10 Bloomsbury Street
London WC1B 3SR
England

For my wife Charlie

Contents

Figures

PROLOGUE

Enter a virus stage-East . . .

Timeline: Early March 2020. London. World Health Organization declares COVID-19 a pandemic. Global COVID-19 cases 87,137 (mostly in China, some on a boat); deaths 2,977.

*

NATURE WANTS YOU dead. Not just you, but your children and everyone you have ever met and everyone they have ever met; in fact, everyone. It wants you to cough and sneeze and poop yourself into an early grave. It wants your blood vessels to burst and pustules to explode all over your body. And until relatively recently it was really good at doing this. In 1900, the average life expectancy of a human was thirty-one years. Really, I should be dead already.

But thanks to advances in scientific understanding, better hygiene and the massive impact of modern medicine, I am not dead – and neither are you. We owe our extended years to the scientists who discovered, invented and tested new approaches to control infectious disease. Drawing on my

* Each represents one million cases of COVID-19.

twenty-five years' experience as a researcher, I will explore how human ingenuity has led to amazing breakthroughs in controlling, preventing and treating infectious disease.

This book will describe the personalities behind the breakthroughs. Some individuals have become synonymous with the identification and prevention of the micro-organisms that cause infectious diseases: Alexander Fleming, Jonas Salk, Edward Jenner, Florence Nightingale, Louis Pasteur, Robert Koch and latterly Tony Fauci. Yet there are also lots of less well-known scientists, often uncelebrated, who were critical in getting us to where we are today: Maurice Hilleman, Félix d'Hérelle, Alice Ball, Lady Mary Wortley Montagu, Françoise Barré-Sinoussi, Tu Youyou, Dorothy Hodgkin and John Snow. I have tried to redress the balance by highlighting some of the incredible scientists you haven't heard of.

For every slightly famous scientist mentioned an additional cast of thousands go nameless, without whom we would probably still be holding posies to our noses and taking purgatives to prevent disease. It would be impossible to namecheck everyone who has ever wielded a pipette in the fight against infectious disease. However, where possible, I will try and interweave the work done by my team as a kind of paean to the unknown scientist.*

One thing I discovered while researching the history of humanity's war on bugs is that the behaviour of the superstars has not always been exemplary. Admittedly, some of this is seen through the lens of twenty-first-century values; whether it is OK or not to apply today's moral framework

* Also, if you can't describe your own research then what's the point of writing a book?

to past events is probably outside the scope of this book. Either way, some famous scientists were clearly terrible human beings, regardless of whether they are held up to contemporaneous or current day standards. And this raises an important question – is it better to do something great but be a terrible human or be a good human but do nothing exceptional? I don't know. One thing is for certain: there are quite a few scientists who are both terrible human beings AND have done nothing exceptional.

This book is very much a product of the pandemic. I have tried to reflect this with a timeline of when each chapter was written, where I was when writing it and the global scale of the outbreak at that time. I started writing in late March 2020, when my research lab at Imperial College London locked down in accordance with UK government guidelines. Before then, I had little desire or motivation to write a book. I was quite happy in my world of Petri dishes, viral cultures and somewhat esoteric science, chipping away at the mysteries of infection. But then there was suddenly a huge demand for what I and other scientists knew. Admittedly, I was a bit surprised that most people didn't already carry this random stuff in their heads – who doesn't need to know the record amount of rice-water stool produced, or the length of the longest tapeworm, or the binding receptor for the influenza virus?

The demand for information came at a personal level – with friends asking for advice about testing, vaccines or shielding. It came at a national level – as one of a wider community of scientists, I received over three thousand emails in 2020 from journalists asking for commentary, interviews and explanations of the rapidly evolving situation; from debunking nonsense to explaining how vaccines work. And it came at a political level;

during the pandemic, politics and science became much closer bedfellows. Many of my colleagues at Imperial College London (Wendy Barclay, Steven Riley, Neil Ferguson, Nick Grassly, Paul Kellam, Peter Openshaw and Alison Holmes, to name but a few) were amongst the scientists who contributed their time and knowledge to the Scientific Advisory Group for Emergencies (SAGE) and its various subcommittees, to help direct the response to the growing crisis.

The pandemic drove home the need for a better understanding of infectious disease biology at all levels, because of its huge impact on every aspect of our lives. In response to this hunger for information, I have tried to distil our understanding of infectious disease in order to put the COVID-19 pandemic into the context of humanity's progress.

And so, speaking of context, let's rewind one hundred years . . .

* * *

Viruses have always had the potential to cause mass disruption. In 1918, in the aftermath of the First World War, an influenza pandemic (sometimes called Spanish flu) raged around the world, killing fifty to one hundred million people, far outstripping the death toll of the war itself. After it had subsided, *The Times* referred to the pandemic as 'a hurricane across the green fields of life, sweeping away our youth in hundreds of thousands'.[1]

Just over a century later, at the Wuhan South China Seafood Wholesale Market (probably), a novel virus emerged, SARS-CoV-2. From the initial starting point in Wuhan, China, it seeded clusters of infection in ski resorts in Italy and shrines in Iran that rapidly disseminated to other countries, driving the COVID-19 pandemic.

And SARS-CoV 2 is only one of a multitude of pathogens that threaten human health. The COVID-19 pandemic can't even claim to be the first major viral outbreak of the twenty-first century – humanity survived SARS in 2003, swine flu in 2009 and Ebola in 2013.

And pandemic viruses are only the tip of the infection iceberg. Many pathogens are endemic, ticking over in the population and leading to disease and death. In 2019, the World Health Organization (the WHO, pronounced phonetically W-H-O) recorded 5 million deaths from infectious disease, including 1.2 million from TB, 675,000 from HIV/ AIDS, 410,000 from malaria, 165,000 from measles, 110,000 from whooping cough and 47,000 from tetanus. The last three are particularly tragic because of the widespread availability of effective vaccines.

And yet, infectious disease is much less likely to kill you than non-infectious disease. It is a testament to science's success in controlling infectious disease that most of us will die of non-infectious diseases – even during a full-blown pandemic. In England and Wales there were estimated to be 604,045 deaths in 2020 (Office for National Statistics figures[2]), of which 77,686 had COVID-19 on the death certificate. Each of which was a tragedy. Yet 90% of the deaths came from other causes, including dementia, diabetes and heart attack.

But 2020 was in no way a normal year, because in most years infectious disease doesn't even get a look-in. In 2017, the top five causes of death in the UK for people under eighty were all non-infectious. Influenza and pneumonia only entered the top five in people aged over eighty. In 2019, infection caused only two of the top ten worldwide causes of death (respiratory tract infections and diarrhoea).[3] In that year, more people died from road traffic accidents than either TB or diarrhoea. The picture changes in lower-income countries, where more deaths are due to infections because of poorer healthcare and unsanitary conditions.

WINNING THE FIGHT

The central theme of this book is that, yes, infectious diseases *can* be incredibly disruptive and destructive, but we now possess the tools to put them back in their box; and at speed. The COVID-19 pandemic brought into sharp focus the role of modern medicine in the control of infectious disease – low-tech (non-pharmaceutical) interventions, diagnostics, antiviral drugs and vaccines.

Understanding the timeline of the COVID-19 pandemic gives insight into the extraordinary speed of the response (see figure 1). In November 2019, a cluster of pneumonia cases with no known cause was detected in Wuhan, China. Working backwards, the first case probably occurred around 17 November 2019. In December 2019, the Chinese Centre for Disease Control started an investigation into the spread of this novel pneumonia.[4] By the end of December the Chinese team had collected samples from the lungs of infected patients and started trying to identify the cause. They discovered a new virus that had some relationship to a bat coronavirus and they shared the genetic sequence of the virus publicly on 10 January 2020, fifty-four days after the first case.

This speed is staggering. Even more remarkably, the first doses of vaccines for clinical trials were ready on 7 February 2020, twenty-eight days later. The incredibly fast timescale of interventions during the COVID-19 outbreak reflects the distance we have come.

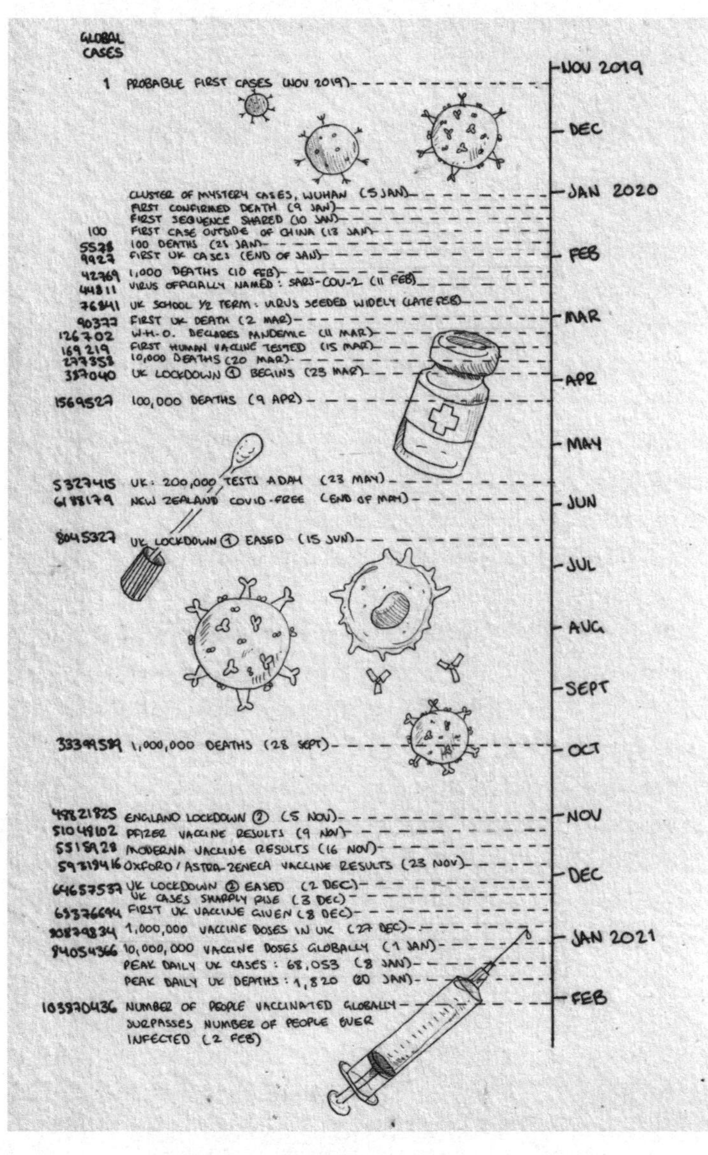

Figure 1 Timeline of COVID-19: Key events between the first case in November 2019 and February 2021.

Not convinced? Smallpox gives us a useful comparator to see just how fast the progress with SARS-CoV-2 was. Smallpox disease has been a scourge since at least the time of the pharaohs (evidence suggests that Ramses V was infected). Yet for most of its history no one knew the cause of smallpox. Variola virus, the causative agent, was only identified in 1935 (by Magarinos Torres and José de Castro Teixeira at the Oswaldo Cruz Institute in Brazil). Researchers identified SARS-CoV-2 nearly forty thousand times quicker, taking weeks not millennia. From the first genome sequence published to the first experimental vaccine dose designed, manufactured and administered was seventy-one days, less time than it takes to make cheese. From the first recorded case to the first licensed vaccine injection took less than one year. That is mind-blowing: you can wait longer for tickets to *Hamilton*.

Still not sure about our progress? How about a comparison with 1918 flu? One of the most remarkable things to me is that the causative agent of the 1918 pandemic was unknown at the time. Little had changed since the eighteenth century when the name influenza, which comes from the astral influences that were then presumed to cause it, first entered the English language. As late as 1932, the textbooks still attributed influenza disease to a bacterial infection, *Haemophilus influenzae*, rather than a viral one.[5]

This accelerated timeline of discovery, from mystery disease to causative agent, didn't happen by accident – it was the result of decades of research. Immense strides in technology underpin infectious disease research, with advances in one field leading to benefits in another.

Huge global pandemics aside, things are definitely better than we might think. In *Factfulness*, Hans Rosling makes a compelling data-based case for the secret silent miracle of human progress. Much of this progress revolves around infection.[6] The story of infectious diseases has an overwhelmingly positive message – that through expertise and ingenuity we have overcome pathogens that historically had a huge impact on humanity. As a result of science's success, even during a full-blown pandemic most of us will die of non-infectious causes, in stark contrast to a hundred years ago, when many more people died of infections than other causes. The tipping point came somewhere in the 1950s, when the cumulative effect of access to clean water and increased vaccine coverage changed our relationship with microorganisms.

This book will celebrate these successes and explain the underpinning science, as well as looking into background stories and research that led to these advances. The global nature of infection is a recurring theme. Unless you were the person to hunt the monkey or eat the bat, you will have caught an infection from another person and will probably pass it on to someone else. Therefore all of us must play our part in the control of infections.

Part 1 of the book will cover the 'ologies – microbiology, virology, immunology and epidemiology; the underpinning science of infection and how the body fights it. It will also cover the tools used to diagnose the infection-causing agents. Part 2 of the book will look at prevention and cure, starting with low-tech solutions that can be effective in controlling spread, the use of vaccines to prevent infection happening in the first place and the drugs to treat infections. Finally, we will step back and look at how infectious disease research both shapes and is shaped by the human context, before taking a wild guess about future technologies.

But first of all it's vital to understand what pathogens are and how and why they infect us. The first chapter of the book will tell the story of the discovery of pathogens and how we came to understand their role in infectious disease. And where better to start than the beginning of life on earth?

PART 1

'Ologies: Investigating and Understanding Infectious Disease

CHAPTER 1

A Brief History of Microbiology

Timeline: Late March 2020. Epsom. COVID-19 spreading globally, one week into UK lockdown. Global COVID-19 cases 509,164; deaths 23,335.

'Begin at the beginning,' the King said gravely, 'and go on till you come to the end: then stop.'

Lewis Carroll

IN THE BEGINNING was the word and the word was soup. And not a good soup like chicken noodle or mulligatawny, but rather a primordial soup. This soup contained some really long molecules, a number of which evolved to make copies of themselves. These self-replicating molecules developed further and over time evolved into organisms demonstrating the characteristics of life: they could move, respire, respond to the environment, grow, reproduce, excrete and eat. More time passed and things got more complicated. Which is roughly where we come in as a species. *Homo sapiens* have walked the earth for approximately 200,000 years – an extremely long time in human terms, but nothing at all in evolutionary terms.

So we find ourselves something of a latecomer to the scene. This means we had to adapt to our surroundings to get the resources we needed to make copies of ourselves. In evolutionary biology this is called a niche. Different organisms have developed different tactics to acquire resources. Some trap energy from the sun (plants mostly), some eat plants (herbivores) and some eat herbivores (carnivores) – with a variety of strategies in between. So far, so primary school.

Critically, resources are limited, leading to competition both between and within species. Competition for resources drives evolution. When you get down to brass tacks, evolution is about the struggle to pass your genes from one generation to another. All living organisms, except maybe incels, are trying to achieve this one goal. If you are sufficiently curious, read Richard Dawkins' *The Selfish Gene*, still the definitive popular science primer on evolution.[1]

Some organisms preferred to hitch a ride on their hardworking relatives. Rather than go to the effort of finding ways to make copies of themselves, they evolved methods of invading other cells, getting them to do the hard work. This is called parasitism. Not all parasites are microorganisms – mistletoe, for example, steals its nutrients from the plant upon which it grows. However, this book focuses on the microbiological level: specifically, how do tiny parasitic organisms threaten our whole body?

Time for a bit of terminology. Scientists are sticklers for the correct word because it avoids ambiguity. A disease is not the same thing as the thing that causes it. Disease refers to the symptoms. So COVID-19 is the un-catchy name for the disease caused by the similarly un-catchy SARS-CoV-2 virus. Disease can be infectious or non-infectious. Non-infectious diseases are also known as non-communicable diseases because you can't catch them from someone else – common examples include heart disease, lung cancer and stroke. You get non-communicable disease through a combination of environment, genes and bad luck. Infectious diseases are caused by another organism invading your body.

The catch-all term for microbes that cause infectious diseases is pathogens. Sometimes people refer to them as bugs. They shouldn't. Bugs live in the garden. So, from now on, call them pathogens. Pathogens are anything (living and small) that can get into your body and cause disease.

LIVING TOGETHER: DYING TOGETHER

Our interactions with pathogens have shaped us. They have shaped our societies, our genes and our very selves.

Pathogens have had a huge influence on human history, all the way back to prehistory when pathogens like smallpox spread from the first farm animals to the earliest farmers. This interspecies movement is called zoonosis and it still occurs in the present day. HIV most likely arose in monkeys and SARS-CoV-2 probably came from bats. But before getting on your high horse about not eating wild animals or bush meat, bear in mind that influenza, which kills approximately 500,000 people *every* year, comes from domesticated chickens and pigs.

While infections have coexisted with humans throughout our evolution, pandemics are a feature of civilisation. The word pandemic comes from the Greek, meaning *all* the people. People need to be living in close enough proximity to allow the infection to spread within communities and have enough contact between communities to spread it more broadly. The first recorded pandemic was the Justinian plague. This emerged in Egypt in the year 541 CE and did not disappear until 750 CE. Pandemics accelerate societal change. The Justinian plague led to the destabilisation of the Roman Empire, creating space for other powers to emerge, including Islam. The Black Death, caused by the same bacteria as the Justinian plague, *Yersinia pestis*, had an even greater societal impact, killing a third of the population of Europe between 1347 and 1351.*

Y. pestis is just one of several infections that have altered the course of human history. The smallpox virus enabled five hundred conquistadores to annex the Aztec Empire, with a little help from their guns and horses.[2] In 1520, a single pox-ridden slave from Spanish Cuba arrived in Mexico; the virus they carried cut the indigenous population from 20 million to 1.6 million by 1618.

Infections have not only shaped our societies; they have also shaped our genetics. The need to survive infections puts a selective pressure on certain characteristics. For example, the higher prevalence of malaria in Africa explains the relative frequency of the genetic condition sickle cell anaemia in people from Africa. Sickle cell disease arises from a mutation in the haemoglobin gene, which affects how people make red blood cells. Without the mutation red blood cells look like fried eggs and with the mutation they look like bananas (or

* A quick note on the use of bacteria as both singular and plural. Some people (as well as the inbuilt spellchecker in MS Word) insist on bacterium being used for the singular. They are the same people who use datum as a singular for data. They are probably right, but since this is my book I will use neither. If you feel obliged to use the word bacterium, either write your own book or change it with a pencil.

sickles to be more accurate). In a world without malaria the sickle variant is a disadvantage, because the banana-shaped red cells do not flow around the body as efficiently. However, the *Plasmodium* parasite that causes malaria prefers fried eggs (healthy red blood cells) to bananas and so people with the sickle haemoglobin gene find themselves less susceptible to infection. Since malaria is more common in sub-Saharan Africa, individuals who don't get infected are more likely to survive and pass on their genes to the next generation.

There are many other valid examples, but sadly I think I dreamed up the other example I wanted to use – linking a preference for tea rather than beer in the Middle Ages to the selection for the alcohol flush reaction in people of East Asian origin. If anything, evidence suggests that mutations in the gene involved in ethanol metabolism (alcohol dehydrogenase), which is associated with alcohol flush, emerged with the development of rice wine.[3]

Lastly, at a cellular level, humans are the offspring of host and microbe. The energy required by my brain to type this and for yours to read it is produced by a part of the cell called the mitochondria. Mitochondria are derived from bacteria in an ancient fusion event. As an aside, we only inherit our mitochondria from our mothers – hence all humanity can be traced back to a single ancestor, known as Eve, who lived in Africa 150,000 years ago. It's not just the machinery our cells use that is bacterially derived – 8% of our DNA is microbial in origin. Basically, some viruses got lazy, set up homes in our cells and never left. These viruses, called the endogenous retroviruses, have been part of our make-up for so long we actually need their genes to function. The gene Syncytin-1, which helps the placenta develop by fusing cells together into one larger cell, is of viral origin.

It has been part of our genomes for about twenty-five million years – longer than humanity's existence as a species.

In turn, we have shaped pathogens. There is an ongoing arms race between host and infection, where both change but neither wins. The US evolutionary biologist Leigh Van Valen coined the Red Queen hypothesis to describe this race, based on the line in *Through the Looking-Glass* where the Red Queen says: 'Now, here, you see, it takes all the running you can do, to keep in the same place.'

Even though infections have had a great impact on us and us on them, for most of history humans remained ignorant as to why they got sick and died. All sorts of things were blamed for illness, like bad air, the stars, cats, ethnic minorities and witchcraft. Until the eighteenth century the Greek philosophers Hippocrates and Galen influenced cutting-edge scientific thinking about infections in Europe, attributing disease to a misbalance of the four humours – phlegm, black bile, yellow bile and blood. The lack of understanding led to terrible treatments you would find only on quack websites today: bleeding, trepanning and purgatives. It was the survival of the luckiest and your health was probably improved by avoiding a doctor in the first place.

It took a long time to realise microorganisms caused infections. Yes, there was speculation about tiny organisms early in the sixth century, but given some of the other things that were believed then this feels like an infinite monkeys and typewriters situation; if you look hard enough, you can find someone who proposed the answer by accident without any evidence to support it.

Scientific progress and technological advances have always gone hand in hand: better tools lead to better science.

Antonie van Leeuwenhoek, a seventeenth-century Dutch draper, refocused our understanding of microorganisms. He initially wanted to investigate the quality of the thread he sold, which led to an interest in lens making and ultimately the development of the microscope. Van Leeuwenhoek's microscope enabled him to see organisms invisible to the naked eye. He observed a whole range of different single cells in blood, pond water and semen (because you would, wouldn't you).

The demonstration by van Leeuwenhoek that single-celled organisms existed unlocked understanding about their role in disease. Two Germans and a Frenchman, Cohn, Koch and Pasteur, built on this knowledge to lay the foundations of bacteriology in the nineteenth century. Ferdinand Cohn worked at Breslau University (now Wrocław, Poland), where he used a microscope that had been given to him by his father. Being the only researcher at the university to own a microscope gave Cohn a considerable advantage – it would be like turning up to university today with a particle accelerator and a super-computer. Using his unique instrument, Cohn was the first to identify bacteria as a separate class of organism.

Louis Pasteur, he of 'milk is the fastest drink' fame (for this and other terrible loosely scientific dad jokes see footnote),* was a French polymath who initially started as a chemist, exploring the shapes of chemical molecules. He discovered a property called chirality, which means that individual molecules of the same substance can fold into

* Milk is the fastest drink because it is past-your-eyes before you drink it. See also scones being the fastest food and Dublin the biggest city.

different mirror image conformations. These different conformations of the same molecule are described as left- and right-handed forms – technically L (laevo – left) and D (dextro – right). Like your hands, chiral molecules reflect each other but cannot be overlaid. Chirality has an incredible impact on drug design: just as only a right hand will fit into a right glove, only some drug orientations will match their target. This means that when you make a drug you need to be careful which orientation you get. For example, the drug thalidomide folds into both left- and right-handed structures: the left-handed molecule works for morning sickness relief (as designed) and the right-handed molecule is highly toxic to unborn children.

Having solved the problem of chirality, Pasteur then switched fields to fermentation. He discovered that while some microorganisms cause fermentation of fruit juice into alcohol, other microorganisms can spoil it (for reasons unrecorded he worked on beetroot wine, which one can only imagine is pretty revolting even when not spoiled). Pasteur showed that infections do not just occur spontaneously; they need to be seeded from somewhere else. He proved this by sterilising a nutrient broth in a sealed flask – nothing would grow until it was exposed to the air. He applied this knowledge to remove bacteria from milk to reduce spoilage – in the pasteurisation process.

Robert Koch completes the microbiological triumvirate. Koch worked as a surgeon in the Franco-Prussian War before settling in Berlin as a researcher. He proved that specific microorganisms caused specific diseases. One of Koch's key contributions was a set of postulates (or rules) that can be applied to all infectious diseases. Koch's postulates are still a route map to identifying novel infections. They are as follows:

- First, the infectious microorganism must be found in people with the disease;
- Second, it must be possible to isolate the microorganism;
- Third, the microorganism must cause disease when introduced into a healthy individual;
- Finally, it must be possible to re-isolate the same organism from the newly infected patient.

If you apply these postulates to COVID-19, you can see that the virus SARS-CoV-2 *is* the causative agent and *not* 5G phone masts. SARS-CoV-2 is found in people with COVID-19, the virus can be isolated and grown and if you transfer it to another person you can cause the same disease and recover fresh virus. The simplest refutation of the phone mast theory is the presence of cases of COVID-19 in countries without 5G; the crackpot theory falls at the first postulate.

However, in spite of major breakthroughs in understanding bacteria, in the early twentieth century there were still diseases with no known cause. This was because viruses caused them and viruses remained undiscovered until the 1890s. Curiously, the first virus to be discovered didn't infect humans or even animals: it infected plants, specifically tobacco plants. This infection, tobacco mosaic virus, causes mosaics in the leaves of tobacco. It must be said that when it comes to naming, virologists are extremely literal.

In 1892, Dmitri Ivanovsky, a Russian botanist, was investigating damage to Ukrainian tobacco plantations when he discovered that you could transfer the mosaic disease between plants by injecting the sap from an infected plant into an uninfected plant (thus fulfilling one of Koch's postulates). Strikingly to Ivanovsky, when you passed the sap from a diseased plant through a filter so fine it would trap any bacteria (called a Chamberland filter), it remained infectious. This meant that the disease-causing agent was smaller than a bacterial cell. This was the first demonstration of a virus as an infectious agent.

However, it took another six years until a second botanist, Martinus Beijerinck, replicated the study and applied the name 'virus' to the filtered extract. Up until that point the term 'virus' had more generically referred to a poisonous substance. Using a similar approach, but filtering material from a sick animal to infect a new animal, the first animal virus discovered was foot-and-mouth disease. The first human infection to be identified as viral was yellow fever virus in 1901. This led to a cascade of viral discovery through the early twentieth century, identifying causes for diseases of unknown origins including smallpox, polio, herpes, rabies and influenza.

While the existence of viruses could be inferred from the early filtration studies, it was not until 1939 that Helmut Ruska visualised the first virus, using the electron microscope his brother Ernst had invented and he had developed. This was as big a step forward as that made by van Leeuwenhoek two hundred years earlier. Ernst Ruska realised that if you used electrons rather than (visible) light to illuminate the sample, you could get down to a much finer level of resolution. Because the Ruskas worked in Germany during the 1930s and '40s, their work did not become widely known at the time. The record was set straight after the war and in 1986 Ernst Ruska shared the Nobel Prize for his 'fundamental work in electron optics, and for the design of the first electron microscope'.

So, from rather humble beginnings – a soup of chemicals – life on earth evolved into a complex interconnected web of organisms. Antonie van Leeuwenhoek, when he wasn't looking at his own semen down a microscope, pulled the curtain back on a microbial world. Since their discovery, knowledge has exponentially increased about pathogenic microbes, down to the minutest level – understanding how single changes in the genetic code can impact the shape and infectivity of an organism. In order to understand how we can fight pathogens, we need to know a bit more about them; how they infect us and what they do once they enter our bodies. In Chapter 2, I will give an overview of the main classes of pathogens.

CHAPTER 2

The Science of Microbiology

Timeline: Early April 2020. Epsom (should have been at a wedding in Poland), lockdown in UK continues. Global COVID-19 cases 1,914,916; deaths 123,010.

'If you don't like bacteria, you're on the wrong planet. This is the planet of the bacteria.'

Craig Venter

MOST MICROORGANISMS LIVE and die without ever bothering humanity. However, some of them are parasites and can only get the resources they need from us. These microscopic pathogens are the ones of interest here. We will now explore the science of the microorganisms that cause disease: the viruses, bacteria, parasites and fungi (see figure 2).

TAXONOMY: THE NAMING OF PARTS

To help us understand and classify them, we group pathogens into a number of larger families and then smaller families, sub-families, micro-families, strains and isolates. This

Virus

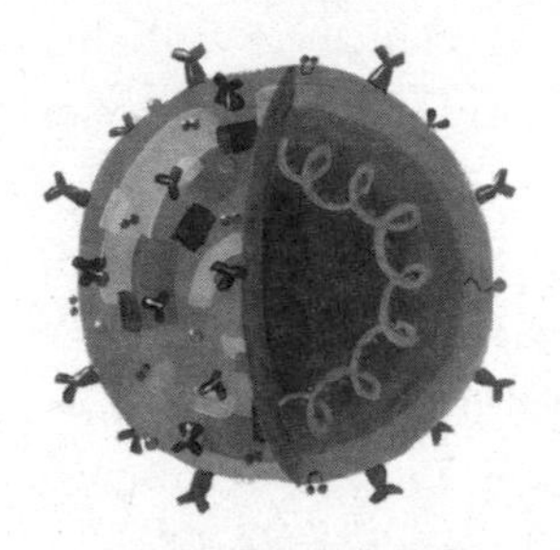

Obligate parasite, Nucleic Acid in a protein coat. Has to infect cell to replicate. **Examples:** HIV, SARS, Influenza

Bacteria

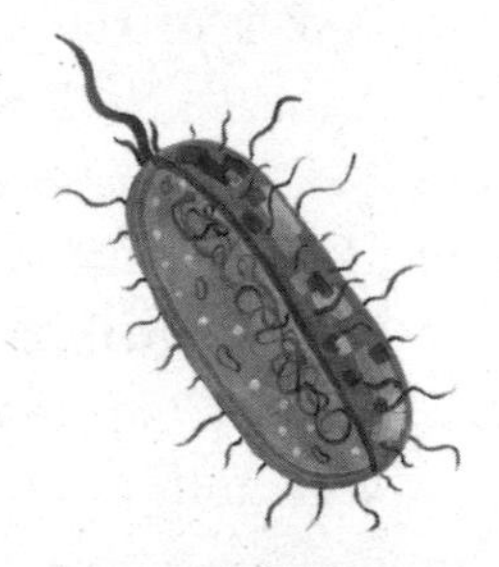

Prokaryote. Very diverse family. Can live independently. Treated with antibiotics. **Examples:** TB, *Staph aureus*, *Salmonella*

Fungi

Eukaryote. Mostly harmless, unless imunocompromised. **Examples:** Athlete's foot, Thrush, *Aspergillus*

Parasite

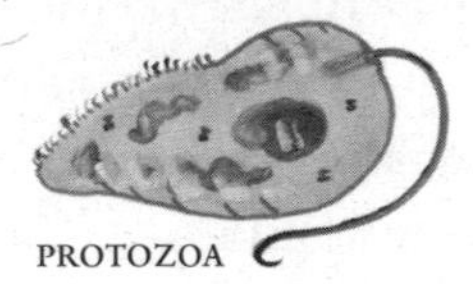

Worms or protozoa. Complex, multi-stage life cycles. **Examples:** Malaria, Leishmaniasis, Ascaris

Figure 2 A pathogen spotter's guide: The four main classes of pathogens are **viruses, bacteria, fungi** and **parasites.**

process of grouping living things is called taxonomy, building on the work of the eighteenth-century Swedish botanist and zoologist Carl Linnaeus (who occasionally referred to himself as *Carolus* in Latin – in a similar way, I occasionally ask my students to call me *Johnus*). Linnaeus introduced a hierarchical system for grouping the natural world, dividing it into three kingdoms: *Regnum Animale*, *Regnum Vegetabile* and *Regnum Lapideum* (hence the game 'animal, vegetable, mineral'). He then further subdivided it into classes, orders, genera then species. Humans are Animals-Chordates-Mammals-Primates-Hominids-*Homo-sapiens*.

As part of his taxonomy, Linnaeus formalised the binomial (two-part) Latin naming system for organisms. In this system the first word reflects the genus the organism belongs to – as an example, the black rat belongs to the genus *Rattus*. The second word refers to the species, which in the example of the black rat is also *rattus*, giving it the official name of *Rattus rattus*. So good they named it twice. Scientific convention is to use the full name the first time a species appears in a text, but after that the genus can be shortened to an initial, giving us *R. rattus*. Brilliantly, there is a scientific name for species whose genus and species are the same – tautonym. Examples include *Giraffa giraffa* (the giraffe), *Glis glis* (the edible dormouse) and *Bison bison* (the African tree shrew . . . only kidding, it's the bison).

Given it was written in 1735, the Linnaean taxonomy remains a remarkably robust method of grouping things. It still forms the basis upon which we classify all living things, although the mineral kingdom is no longer used. Because the living world is substantially more complicated than Linnaeus understood it to be, there has been an ongoing restructuring of his hierarchy. To encompass the huge diversity of life, in 1990 Carl Woese added an additional level above kingdom: the domain. Pathogens by and large fall into two of these domains: the bacteria and the eukaryotes (pronounced you-carry-oats).* Eukaryotes include animals, plants and fungi. It goes without saying that nothing in biology fits neatly and, alas, viruses don't really fit into one domain. This is because, according to the people who are into this kind of thing, they are not alive. Except they are. Sort of. At least you can kill them. Sort of.

GENETICS 101

All pathogens are trying to steal your resources so they can make copies of themselves and pass their genes on to the next generation. To understand the battle between human and pathogen, you need to understand a bit of genetics (see figure 3). The following explanation would make the genetics lecturers who taught me pull their hair out, but it is close enough. Genes are the units of heredity: they are the instructions for what makes you you and not, say, a blobfish (*Psychrolutes marcidus*, the world's ugliest animal). They are passed from parent to child, explaining why your children look a bit like you and you look a bit like your parents.

* One of the unexpected trials of the pandemic lockdown was having to move all of my academic teaching material online. This involved using automated software to transcribe my words into text. The software was not a fan of my name (I became Dr George Conning), or my subject matter, rendering CD4 T cells into seedy for tea sells.

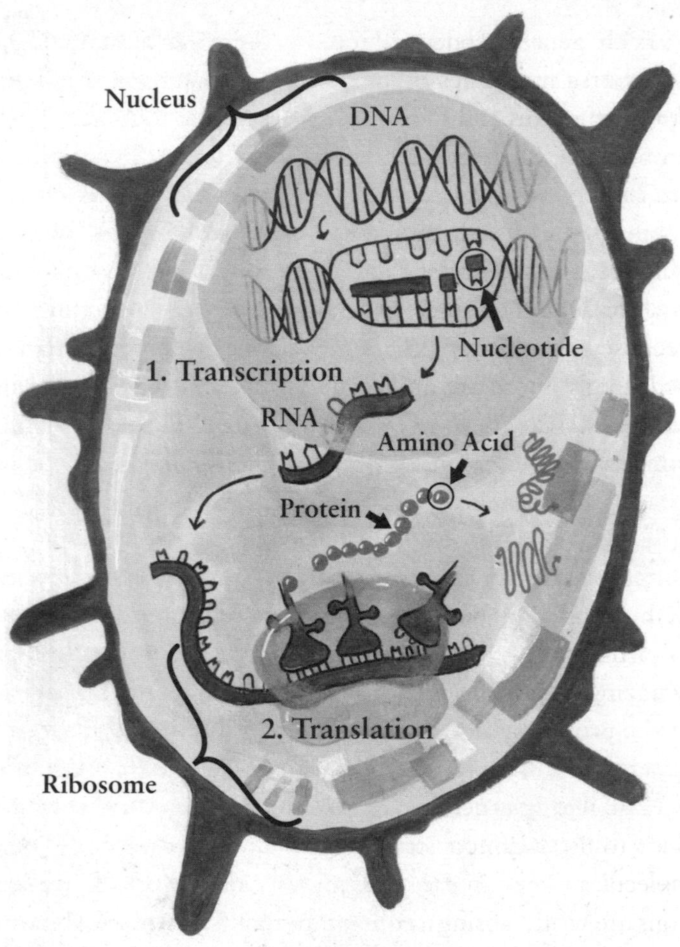

Human Cell

Figure 3 DNA into RNA into proteins: The information is stored on DNA in the cell nucleus. Step 1. **Transcription**. This information is transferred to the protein factories of the cell (the ribosomes) on a different molecule called RNA. DNA into RNA is called transcription. Step 2. **Translation**. The RNA encoded information is translated into protein, which is constructed from amino acid building blocks.

Each gene encodes a single protein. Proteins are the workhorse molecules of the cell – they catalyse reactions, build structures and speak to other proteins. A bit like Lego, proteins are constructed from building blocks. These blocks are called amino acids and there are twenty-one of them. While I won't list them all here, it pleases me to know that tyrosine is named after the Greek word for cheese, asparagine for asparagus and serine after the Greek word for silk, because of the materials from which they were first isolated. Amazingly, there are only twenty-one amino acid building blocks, which doesn't seem like enough. However, in the same way you can build a castle, the Millennium Falcon or a double-decker couch using only twenty-one different types of Lego, from the twenty-one amino acids available the human body can build twenty thousand unique proteins with a huge range of functions, from structural (collagen) to virus sensing (toll-like receptor 7) to digestive (pepsin). Amazingly (or not), we still don't know the function of all of the proteins made by human cells.

The nucleus acts as a central reference library to the rest of the cell. The instructions that tell a human cell which amino acids to use when constructing proteins are written in DNA molecules stored in the cell's nucleus and the DNA instructions are written using a combination of four letters. They are not actual letters, but chemical building blocks called bases: A (adenine), T (thymine), C (cytosine) and G (guanine). The bases are strung together on a double helix molecule, as first identified by Franklin, Watson and Crick. The DNA double helix contains two complementary copies intertwined: A always pairs with T and C with G; these are called base pairs.

The DNA encoded message is converted into proteins in two steps. First, the instructions are copied from the master copy stored in the cell nucleus into a separate molecule called RNA. The process of transferring the instructions from DNA to RNA is called transcription. The transcribed message is then carried from the nucleus to the ribosomes, which behave a bit like 3D printers. They can make whatever shape they are instructed to by the RNA. The printing of protein from RNA instructions is called translation.

How can all of life be encoded in just four letters? The key is that complexity can be built through repeated patterns: a bit like computers turning a string of zeros and ones into pictures of Kim Kardashian's bottom. In DNA, the letters are grouped into triads and each of these groups of three represents a specific amino acid; for example, the triad ATG encodes the amino acid methionine. This is called the genetic code. Since there are sixty-four three letter combinations using A, T, G and C (try it yourself at home: AAA, AAT, AAC, AAG, etc.) and only twenty-one amino acids, there is some redundancy in the genetic code; for example, the amino acid phenylalanine is encoded by either TTT or TTC in DNA. The ribosomes read along the string of RNA like an old ticker-tape reader, adding the correct amino acid for every three bases read. One other important feature of the genetic code is that three of the combinations (TAA, TAG and TGA) tell the ribosomes to stop, ensuring proteins are not infinitely long.

While it is useful to know the above, you need to understand two key messages to make sense of the rest of the story: all organisms use the same building blocks of life and DNA can change.

- Shared building blocks: Amino acids and the constituent bases of DNA cycle through all living things. The proteins that make the lens of your eye, enabling you to read this, were once part of the cow in the burger you just ate, and before that were part of a blade of grass. In turn, pathogens want to steal your amino acids from you.

- Mutation: The message encoded in the DNA is not set in stone forever: over time it can mutate. Most mutations lead to nonsense – because who would put an aspartate after an alanine residue at position three?* Some mutations make no difference and others make things worse, causing cancer and genetic disease. But a minuscule number improve the function of the protein. Mutation helps pathogens escape from the immune response and adapt to drugs.

GOING VIRAL

Armed with your newly acquired knowledge of taxonomy and genetics, let's now look at the four types of pathogens that cause most human infections. This is an overview and I will go into more details later about bacteria (see Chapter 10), viruses (see Chapter 11), fungi (see Chapter 12) and parasites (see Chapter 13) when talking about drugs to control them. But where better to start than viruses?

Viruses are ubiquitous, existing everywhere, infecting everything. Viruses infect plants, insects and bacteria: basically any living thing. Viruses are obligate parasites. Most other organisms can make copies of themselves independently; viruses can only reproduce if they invade other cells. They are masters of invasion, evolving to be professional cell hijackers. All the time we've been evolving opposable thumbs and large brains, viruses have been developing methods to invade cells.

* A little genetics humour for you there.

Viruses can be named after the disease they cause: influenza virus is named after influenza and yellow fever virus after yellow fever. This is admittedly confusing but it reflects how the disease was known before the causative agent. Alternatively, viruses are named after the part of the body they infect, in Greek to make it sound more sciencey; hence rhinovirus rather than nose virus (*rhino* is the Greek word for nose). Finally, viruses have been named after the geographical region where they were discovered (Ebola after the Ebola river and Lassa after a village in Nigeria). The geographical naming of viruses stopped due to the stigma attached (which is why SARS-CoV-2 took three months to be named and wasn't called Wuhan virus), though it was initially used for emerging genetic variants (e.g. the Kent, now Alpha, strain).

It's at this point I would like to introduce a major character – respiratory syncytial virus or RSV. It may not be as showy as influenza or SARS-CoV-2 but it does cause a significant level of disease. It is also the virus that I have spent a lot of my career working on, ever since starting my postdoctoral research with Peter Openshaw in 2003. It was named RSV because it is a respiratory virus which causes infected cells to fuse together to form syncytia (did I mention how unpoetic virologists are?).

RSV is extremely common in children. All children will be infected with the virus in their first two years of life – most cases are mild but 3–5% of all children infected with RSV will need hospitalisation. At the peak of the RSV season (which used to be October to March in the UK) infection accounts for one in six of all UK paediatric hospitalisations.

In 2008 this included my son. We were on our way to Cornwall for a family holiday. Our son had a bit of a concerning cough as we left the house, but it didn't seem much more than standard baby stuff to us. Fortunately, we stopped overnight with friends (Drs Laura and Rob Bethune) and they rushed us to Taunton hospital. Where we stayed for a week. It wasn't quite the first family holiday we had imagined. I had been working on RSV for about five years before I had my son and knew the numbers of cases, but it was always a bit academic until I was standing outside a paediatric intensive care unit crying uncontrollably after my son recovered. It was a formative moment.

The hospital was kind enough to provide us with a private room, as we found ourselves four hours from home, and apart from the short periods when we could see our son we had little to do. So, and this probably sounds odd, we spent the time rehearsing my talk for an upcoming interview to become a lecturer at St George's University in London. Which, thanks in some part to being more prepared than I have ever been, I got, leading me to where I am today. I like to believe that part of the reason I still work in infectious disease is because of what happened to my son.

In high-income countries like the UK, where there are enough paediatric intensive care beds, most children survive – my son included. However, in low- and middle-income countries (LMIC), with fewer intensive care beds for children, the survival rate is significantly lower and RSV causes nearly 200,000 deaths a year around the world.

I have introduced RSV because after eighteen years of working on it, many of my research anecdotes relate to it. I will now use RSV, influenza and SARS to illustrate key features of all viruses:

1. A genetic message. Every virus contains genetic material itching to make a copy of itself. Viruses are the budget airline versions of life, stripped back to the absolute

minimum number of functions needed to enable them to replicate and no more. This is because replicating genes comes at a cost; each base in a nucleic acid needs to be obtained from somewhere and like all synthesis reactions where new molecules are made, adding bases to the main nucleic acid requires energy.

Viruses are unusual: unlike other organisms, which always use DNA to pass information between generations, viral genetic material isn't always DNA. Some viruses use RNA, which human cells mostly use to transmit information from the nucleus to the machinery that makes proteins. These differences in the way in which viruses encode their genes have a profound impact on viral evolution. When human cells make copies of themselves, they use proofreading to ensure that the DNA copies are identical to the original code – this reduces the rate of mutation.* However, viruses that use RNA to transmit their genes lack this proofreading capacity, leading to a much higher mutation rate. While this can be deleterious for an individual offspring virus, for the population as a whole it is extremely effective. This comes down to numbers: higher species only produce limited numbers of offspring, so each one needs to be as good as possible. Viruses go for safety in numbers, really big numbers. Each infected cell produces approximately ten thousand new viruses. Therefore, there is a huge stock of different offspring, a small number of which will be fitter than their parents. This high degree of mutation means it can be tricky to apply the Linnaean system to group

* Similar to how editors check authors haven't sneaked the word boobies into a footnote.

viruses. We can loosely group viruses on a range of different characteristics or the similarity of their genes, but because they change all the bloody time they are hard to pin down exactly or even name (B.1.617.2/ delta anyone?).

2. A coat. Viral genetic material is somewhat fragile. The virus needs to defend it because other organisms want to destroy it to defend themselves from infection. Viruses therefore package up their genes in a protein coat. The protein coat is important for viral transmission. A combination of factors associated with the coat protein drive viral spread, including what types of cells viruses can infect and how stable viruses are in the environment outside the body. Some viruses, called enveloped viruses, pick up another outer coat of lipids from the cell membrane as they exit the host cell, which can give them an extra layer of protection.
3. Entry proteins. Wrapping your genes in a nice protein coat isn't enough to get you anywhere. To take over a cell's chemicals the virus needs to be able to invade it. Viruses use proteins on the outside of their coat to enter other cells, which affects their appearance. The corona in coronavirus refers to the entry proteins that stick out from the virus, making it look like a crown on an electron microscope. The entry proteins made by SARS-CoV-2, influenza and RSV are called spike, haemagglutinin and fusion respectively. Entry proteins pull the virus and the cell close enough together that the virus can inject its poison.

Viral entry proteins target molecules on the host cells: SARS-CoV-2 targets a molecule on lung cells called the ACE2 receptor, influenza uses sialic acid and RSV nucleolin. The location of the molecule that the virus targets for entry determines which cells the virus infects, because different human cells express different proteins on their cell surfaces. This cellular targeting is called tropism. The cellular targets for RSV and influenza are only found in the lungs, which is why they cause respiratory infections. However, ACE2 is widely distributed throughout the body, which may account for some of the severity of COVID-19. Once it enters the body it can infect many different cells.

As well as determining which cell types a virus infects, the entry proteins are important in interspecies spread. One of the factors that determines whether a virus can jump between different species is the similarity between the virus targets on the cells. SARS-CoV-2 was able to bind to a wide range of ACE2 receptors from several different species, including humans, rabbits, bats and pangolins.

Influenza is another example of a virus that can jump between species. The molecule that influenza binds to, sialic acid, attaches to cells in two different orientations called α2–3 and α2–6, based on the way in which the molecules point. Influenza viruses that infect birds prefer the α2–3 sialic acid orientation and human influenza viruses target the α2–6 sialic acid orientation. Both configurations are found in the human airways, meaning we can potentially be infected with bird flu viruses.

Since the entry proteins appear on the outside of the virus, they are the key feature of the virus recognised by the immune system and because of this vaccines target them. All of the licensed COVID-19 vaccines target the spike protein that sticks out of the SARS-CoV-2 virus and influenza vaccines mostly target haemagglutinin.

4. A capsid. The capsid is a coat within a coat, used by the virus to hide from the immune system. Immune systems are very good at recognising genetic material from other

organisms: it acts as a danger signal, triggering a rapid response that then leads to the body killing the infection (more of which in Chapter 5). Since viruses are trying to sneak their genetic material into the cell, they don't want to trigger the alarm system. One way they avoid the host alarm system is to wrap their genetic material up in a protein, so it isn't seen. There are various degrees of sophistication by which the viruses do this: some just wrap it up like a cotton reel; however, others form structures made of multiple repeats of the same protein, like a football.

5. Genetic replication machinery. Genetic material doesn't just replicate itself. Human cells contain a whole suite of proteins used to make copies of DNA and RNA. DNA replication is a tightly controlled process to avoid the introduction of mistakes and to ensure copies are made only when required. Part of the viral life cycle is stealing the raw ingredients of nucleic acid – the As, Ts, Cs and Gs – and co-opting them and the cellular machinery to make more virus rather than more host. To achieve this, the virus makes proteins that can hijack the host gene replication machinery.
6. Immune evasion. The main threat to viruses is the immune system and viruses invest a lot of effort in hiding from it, in various nefarious ways. All viruses need to escape the immune system in order to infect their hosts. Immune evasion can be loosely separated into avoidance and interference.

Immune evasion by avoidance can be achieved by hiding in places the immune system doesn't normally look – like the eyes (some of the herpes viruses), the nervous system (chickenpox) or the immune system (HIV).

It can also be achieved by mutating the proteins seen by the immune response, necessitating a new flu vaccine each year. As with many of the viruses that infect us, 'influenza' is not a single virus; rather it covers four genera of related viruses – influenza A, B, C and D. The one of highest concern is influenza A. The influenza A genus has multiple subtypes, named according to the proteins on their surface, encoded by the haemagglutinin and neuraminidase gene types. There are eighteen known haemagglutinins and eleven neuraminidases, from which we get the H1N1, H3N2 or H5N1, etc. strains when talking about influenza viruses. Some of the influenza strains mainly infect birds (H5N1 and H7N9), others mainly infect humans (H1N1 and H3N2) and there are those that infect bats (H17N10 and H18N11) – unsurprisingly, because bats, as the zoonosis supervillains, seem to get infected by every going virus. Even within one viral strain, the haemagglutinin protein is not constant and changes over time. All of these changes mean that the immune system fails to recognise the newly altered viruses; a similar situation occurred with the emergence of new variants of SARS-CoV-2 in late 2020 and early 2021.

Immune evasion by interference is where the virus makes proteins that actively target the immune system. Viral interference of the immune system can determine what species of animals the virus can infect. Regardless of how intimate you get with Fluffy, it would be impossible to get cat AIDS, because the feline immunodeficiency virus (FIV) cannot overcome the human cells' defences.

A more concrete example of immune evasion affecting species range can be seen in RSV. RSV has a human version and a cow version. Human RSV cannot infect cows and vice versa, yet swapping a single gene, called NS1, from the cow RSV means the human virus can now infect cattle.[1] For every mechanism the immune system uses to kill viruses, examples of viral proteins that can block it can be found – reflecting the rolling Red Queen conflict between virus and host.

VIRAL DISEASE

Viruses are obligate parasites. They must infect your body to replicate. This means that viruses nearly always cause some form of disease. Because viruses are so diverse, infection leads to a huge range of different outcomes, which we group by infection length and the type of disease the virus causes.

It's about time

Viral infections can be acute, latent or chronic. Some viruses cause short-lived, acute infections; with influenza you kill the virus and recover, or it kills you. The virus banks on you transmitting it to someone else before you kill it. Other viruses (the latent ones) get into your body and stay there hidden as a sleeper agent; chickenpox lingers around and can reactivate later in life, causing shingles. And other viruses (the chronic ones) are engaged in a long-term rolling battle with the body; HIV constantly changes as it seeks to escape the immune system.

Location, location, location

The cells that the virus infects determine the symptoms of viral disease. If a virus replicates in the respiratory tract it will cause colds; if it replicates in the liver it will cause hepatitis; if it replicates in the gut it will cause diarrhoea.

Viral diseases physically manifest in different ways. Some viruses, like Ebola, cause disease by direct cell damage. To replicate and spread, viruses alter the life and death of the cell. Normally, cells live for a finite period after which they undergo a highly controlled process of cell death called apoptosis, which swallows up the contents of the cell. However, this is no good if you are a virus and you want to spread onwards from the infected cell. To achieve greater spread, some viruses induce a different type of cell death called necroptosis. The difference between the two types of cell death mirrors how stars die: apoptosis is like a black hole – the cell collapses in on itself and nothing escapes; whereas necroptosis is like a supernova – the cell explodes, spreading its contents everywhere. Ebola virus causes supernova cell death by bursting the cells lining the blood vessels, leading to the characteristic haemorrhagic fever, with blood leaking from the body.

Cell damage caused by viruses can act as a trigger for an exaggerated immune response; in responding to the virus, the immune system can damage the surrounding cells. A lot of the symptoms we associate with an infection come from the immune system doing its business; for example, fever results from the body raising the temperature to kill infections by cooking them. These symptoms can be beneficial in terms of killing the infection but will also make you feel unwell and in extreme cases the immune response is so strong it actually makes the disease worse. Much of the severe disease associated with SARS-CoV-2 was driven by immunopathology.

Viral infections can also manifest as cancer. Instead of exploding out of the cell, some viruses choose to make it immortal, giving them a nice place to live for ever.* This causes the infected cells to grow out of control; one of the characteristics of cancer. Some cancer-causing viruses, like human papillomavirus (HPV), short-circuit the cell machinery required for controlled cell death. Other viruses cause cancer more accidentally, by inserting themselves into the DNA of the cell in such a way that they trigger a programme of cell replication. Human T-lymphotropic virus (HTLV) infects white blood cells, leading to a form of leukaemia where the patient's white blood cell population becomes dominated by virally infected cells.

BACTERIA

Unlike viruses, bacteria are very much alive. They are just as varied as the viruses. While they might look the same from our perspective, bacteria (or prokaryotes – pronounced pro-carry-oats) are so diverse that the genetic difference between some bacterial families is ten times greater than the difference between humans and trees.

Bacteria have evolved to occupy a wide range of different environments, living in an extraordinary range of conditions, with some bacteria thriving in conditions unconducive to other forms of life – the extremophiles. These include bacteria that can live at crazy temperatures from +120°C in boiling hot thermal vents or be frozen in ice at -20°C, but

* A Shangri-He-La, if you will (this will make more sense later on).

there are also bacteria that can survive acids that would strip the skin from your bones and alkalis that would turn you into soap. All told, they are remarkable.

All this diversity means we need ways to subdivide them. Bacteria are much larger than viruses and are big enough to be seen under the kind of light microscope invented by van Leeuwenhoek. Since they can be seen, they can be grouped by shape, a bit like pasta – spiral ones (spirochetes), round ones (coccus) and rod-shaped ones (bacillus). The shapes are often reflected in the name of the bacteria, e.g. *Streptococcus pneumoniae*. Ferdinand Cohn first identified these shapes in the nineteenth century.

Bacteria can be also subdivided into two large groups by a process called Gram staining, named after a Danish microbiologist called Hans Christian Gram. Gram's stain either colours bacteria purple (Gram positive) or red (Gram negative). The chemical composition of the bacterial cell wall affects how it binds to the dyes. The ability of the cell wall to bind to different chemicals is not only important in distinguishing bacteria but it also influences the resistance to antibiotics, with Gram negatives intrinsically resistant.

Most bacteria in the world are indifferent to humanity, living in soil, water and rotting fish. A subset of bacteria lives in and around humans. Nearly all of these can be loosely described as good bacteria – or, more technically, commensals. These bacteria live in our bodies for mutual benefit. We will look at them in much more detail in Chapter 4, but it is important to know they do exist.

A smaller subset still are the bacteria that can cause disease in humans. Some of these bacteria are misplaced commensals. For example, the bacterial pathogen

Staphylococcus aureus can be found on the skin of a substantial number of people without symptoms. If you took a swab from the skin of a random sample of people, 20% of them (one in five) would have *S. aureus* living happily there. It is not a problem when it lives on the skin; however, when you open up the body – either during surgery or after traumatic injury – *S. aureus* can get inside and cause infections. You may know of it in the context of MRSA (methicillin resistant *Staphylococcus aureus*), which is one of the harbingers of a significant threat to public health, antibiotic-resistant bacteria, which I will return to in Chapter 10.

Other bacteria that can cause disease are predominantly environmental organisms – *Pseudomonas aeruginosa* mostly causes black slime around shower heads and bathroom tiles and it can only invade the airways when your immune system or lung defences are under stress. There are some bacteria that only cause disease when given the space to multiply – *Clostridium difficile* lives in your guts harmlessly and only expands when antibiotics kill other gut bacteria, leading to diarrhoea, fever and dehydration.

Within the disease-causing bacteria, many are happy to live outside cells, funnelling off nutrients as they wash around the blood or lungs; however, there are some that are obligate pathogens, like viruses. These bacteria, often called intracellular bacteria, have evolved so that they must live in human cells. One intracellular bacteria with considerable impact on human history is *Salmonella enterica* serotype Typhi, often referred to as *S.* Typhi. It causes typhoid fever, which led to approximately eighty thousand deaths during the American Civil War. A significant number, given that

200,000 soldiers died from fighting.* The bacterial pathogen that causes the disease *S.* Typhi continues to evolve under human selective pressure and cannot function independently. It is an auxotroph, which means it needs to get some of its key nutrients from external sources (i.e. us).

S. Typhi isn't the only intracellular bacteria that causes disease. *Mycobacterium tuberculosis*, the bacteria that causes tuberculosis (TB), also lives inside human cells, specifically an immune cell called the macrophage. TB has been with humanity for an incredibly long time. It possibly came from cows, or at least it has been around for as long as domesticated cattle. We know this thanks to the ancient Egyptians' weird death rituals: the lungs of 4,000-year-old mummies show evidence of TB-induced damage. Our friend Robert Koch first identified *M. tuberculosis* as the causative agent of TB. Remarkably, he also discovered the causative agents of anthrax (*Bacillus anthracis*) and cholera (*Vibrio cholerae*).

As with the viruses, our understanding of bacteria has come an enormous way since their discovery. As a species, our relationship with bacteria is somewhat different to the viruses, because most bacteria really don't care whether we live or die – while the viruses must infect something. So, while this book focuses on pathogenic bacteria, it is worth remembering that most bacteria are harmless. And chemicals which eradicate all bacteria may do more harm than good.

* Confusingly, there is another bacterial disease that sounds very similar and caused large numbers of deaths in wartime: typhus fever. It is caused by the bacteria *Rickettsia prowazekii* and is spread by lice. In the Napoleonic Wars, typhus fever killed more French soldiers in Russia than Russian cannons or freezing temperatures.

PARASITES

A lot of effort goes into understanding viruses and bacteria, but one of the biggest killers of humanity is neither bacterial nor viral: it is eukaryotic. While the bacterial domain is more genetically diverse, the domain of eukaryotes is much more physically diverse, including humans, sequoia trees, blobfish and malaria.

One of the more eye-opening statistics about infectious disease is the claim that malaria killed half of all the people that ever lived. This remarkable claim was first made in a commentary article in *Nature* in 2002,[2] but with no reference to support it. The claim is probably not quite true. Digging a bit deeper, approximately 108 billion people have ever lived, half of which is 54 billion. That is a lot of people and really quite hard to visualise (though they try here https://www.7billionworld.com/). The peak years for malaria infection occurred at the tail end of the nineteenth century. Extrapolating from the global population at that time (around one billion) and the likelihood of death (one in ten), malaria was causing a hundred million deaths in these peak years. But to get to 54 billion would take 500 years in a row at that level of mortality, which simply isn't the case. Based on population and death rates, Professor Brian Faragher from the Liverpool School of Tropical Medicine estimates that the malarial death total is closer to 5% of everyone who has ever lived. Which is still five billion people, more than half the current population of the world; all told, malaria is a serious problem. In the year 2000, there were 262 million cases of malaria, causing 900,000 deaths. Malaria is caused by a single-celled eukaryotic pathogen from the *Plasmodium* genus. There are five species that can cause malaria, the worst being *P. falciparum*.

Eukaryotic parasites are much more complex than viruses and bacteria. Malaria has a six-stage life cycle, four of which are in humans and two in mosquitoes. When an infected mosquito bites you, it injects the parasite in the sporozoite form and the sporozoites then target the liver cells. These then mature into merozoites, which infect red blood cells. This infection of red blood cells causes most of the disease associated with malaria – the cells become overloaded with parasites and burst, leading to acute anaemia and an inability to transport oxygen around the body. Eventually male and female malaria parasites mate, releasing gametes into the blood, which are taken up by new mosquitoes – and then the cycle restarts. At each stage the malaria parasite looks slightly different and expresses different genes, which makes it a bugger to make drugs and vaccines against.

Malaria is not the only eukaryotic organism that can infect humans. It is part of a larger family of pathogens that are referred to as parasites. This is admittedly quite confusing because pathogenic bacteria and viruses are also parasites. For the sake of this book, I will mostly use parasites as a grouping to mean eukaryotic endoparasites – the ones that live exclusively inside your body. Some parasites, the ectoparasites, live outside the body, but they don't really fall into the infection category and in spite of being interestingly disgusting won't get much coverage in this book: they include leeches, jiggers, fleas, ticks, mosquitoes, bed bugs, lice, scabies, candiru (the Amazonian penis fish) and of course the botfly, which lays eggs under your skin that hatch into maggots (gag).

Eukaryotic parasites can be grouped into two broad categories – protozoa (single-celled like malaria) and helminths (multicellular worms). Protozoa can be further subdivided based on how they move: flagellates use a little propeller-like hair called a flagella; ciliates use lots of tiny hairs to swoosh along; sporozoa don't move; and amoeba move by means of pseudopodia. The pseudopodium, or false foot, is where the amoeba distends its cell membrane into a tube and then the whole organism moves along the tube, which is a bit like squeezing up the last of the toothpaste.

Like a lot of infections, malaria is vector borne – it is transmitted by insect bite. It is not the only pathogen to use another living organism to get into our bodies. Mosquitoes alone transmit many different pathogens: malaria, dengue fever, West Nile virus, Zika virus, chikungunya virus. Mosquitoes are not the only creepy crawlies that can act as disease vectors. You can also get infections from kissing bugs (Chagas disease), tsetse flies (sleeping sickness), sand-flies (leishmaniasis), blackflies (river blindness), snails (schistosomiasis), fleas (plague), waterfleas / *Cyclops* (Guinea worm disease), ticks (Lyme disease), lice (typhus) and chiggers (scrub typhus).

As seen with HIV, Ebola and COVID-19, pathogens also spread from other animal species to humans (zoonoses). Lots of infections can make the species jump – the closer related the species, the more likely the jump. Those of you of a nervous disposition might want to avoid rats (leptospirosis), cats (toxoplasmosis), dogs (rabies), sheep (Q fever), cows (brucellosis), horses (Hendra virus), chickens (campylobacter), pigs (tapeworms), ducks (influenza), lettuce (*E. coli*),* parrots (parrot fever – obviously), chimpanzees (herpes B virus), civet cats (coronavirus), gerbils (ringworm), and even armadillos (leprosy – rarely presumably). And whatever you do avoid

* Not strictly an animal.

bats; they are the mosquitoes of the zoonosis world and can transmit Ebola virus, Marburg virus, Nipah virus, coronavirus, Hendra virus and lyssavirus. Yet even if you avoid all other animals, the species you are most likely to catch something from is the human.

FUNGI

The final family of pathogens to consider are the fungi (pronounced fun-guy, hence the joke). They are everywhere. You can make a sourdough starter because the yeast in the air settles down into the flour and water mix you leave festering in your hipster kitchen. We inhale fungal pathogens with every breath we take. Fortunately for most people we then breathe them back out again or they get trapped in the mucus lining of the airways. However, in immunocompromised individuals – for example patients undergoing chemotherapy – they can establish a colony, leading to severe disease. Don't worry, fungal fans, a fuller exposition of this family comes in Chapter 12 (spoiler alert: tulips play a big role).

Thus endeth the quick introduction to the microbial world around. If I paused from my writing and cast a microscopic eye on everything I can see, I would find a sea of potential pathogens – on my desk, my keyboard, the slice of delicious chocolate cake I am about to eat, in my half-drunk cup of tea. And yet, I am pretty confident I will get to the end of this chapter without sustaining any major infection. Understanding why this doesn't happen is key to preventing infections and treating them when they do occur.

CHAPTER 3

Why Don't We Get Sick?

Timeline: Late April 2020. Still Epsom (should have been at family Easter celebration in Cornwall), still lockdown! Global COVID-19 cases 3,090,445; deaths 217,769.

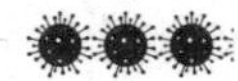

'Breathe, breathe in the air/ Don't be afraid to care'

Roger Waters

I WANT YOU TO take a deep breath. Have you done it? Good.

You just breathed in about half a litre of air.

It is possible to capture the liquid in your breath by cooling it down – a bit like when you breathe on a cold mirror. If you do this, over one minute you can recover 100 microlitres of fluid (about the size of a lentil). If you then grow what you recover from that fluid you get between 100 and 10,000 bacteria, 5,000 viruses and one or two fungi. Every time we breathe, we inhale a cocktail of potential pathogens. And we inhale about 20 times a minute, nearly 30,000 times a day, sucking in 15,000 litres of air.

The world through which we move is awash with potential pathogens. They are not just in the air we breathe; they are in the water we drink, on the lettuce we eat and in the soil on which we stand. And we don't just find microorganisms in the external environment; they are on and in us. Our skin plays host to thousands of bugs and our guts are packed full of bacteria – by some ways of counting there are more cells of bacteria in your body than there are cells of you (more of which later). We very much live in a microbial world. Most of them either ignore our existence or live side by side in perfect harmony. But all of the pathogenic microorganisms – the infections – are just waiting for an in.

Most remarkably, it can take fewer than ten actual viruses to infect you. A virus is 100 nm in diameter,* which means that in a sugar cube made of virus there would be about 1,000,000,000,000,000 viruses, enough to infect every single person on the planet one hundred thousand times.† It was estimated that if you had collected all of the SARS viruses from all of the infected people at the peak of the pandemic, they would not even have filled a Coke can.

Which leads us to the big question: why don't we get sick all the time?

SCIENCE FOR SMARTIES

Before we answer this question, or any scientific question, we need a quick primer in how to do science and what scientific method means for certainty, rare events and crazy theories.

* One nm is one billionth of a metre – see glossary for explanation of science units.

† The sugar cube, the tennis court, the bag of sugar, the London bus and the Eiffel Tower are not standard units, but they are helpful nonetheless.

A key point is that science is a dynamic process: ideas change with increases in information or developments in technology. Each generation is better equipped to interpret the world around them. In the absence of microscopes, which provide evidence of the existence of minuscule living things that can cause infection, linking disease to changes in the season based on observations of prevalence makes as much sense as anything else.

Science is uncertainty, except when we are certain (and even then, we are a bit uncertain). This uncertainty unfortunately opens the door to chancers and snake oil salesmen, especially in the wonderful age of social media (see Chapter 14). The internet has enabled the dissemination of (dis)information on a hitherto unimaginable scale. We saw a large number of semi-plausible and downright idiotic theories during the COVID-19 crisis, many of which just wouldn't die. The most bizarre was the 5G theory, which is straight up wrong, though ingestion of bleach as a cure came a close second.

Some things, though, are more certain than others. If you came looking for subtle sci-comm arguments to persuade you out of your misguided opinions, you are reading the wrong book: the world *is* round, evolution happens, viruses cause diseases. These are not opinions that can be discussed like politics: they are facts.

We establish the facts about infection by using scientific method, the philosophical underpinning of which was developed by Francis Bacon (the sixteenth-century English philosopher not the twentieth-century English painter). The first pillar of scientific methodology is that we build on what went before us. Isaac Newton famously said: 'If I have seen further it is by standing on the shoulders of Giants.' In most instances, our ideas are gradual evolutions and refinements of previous understanding.

There are occasionally larger jumps: Copernicus moving the sun to the centre of the solar system, Newton and gravity, Darwin and evolution, Curie and radioactivity. These jumps are often linked to a single person, but frequently someone else discovered the same thing at the same time. For instance, Alfred Russel Wallace published a paper with Darwin about evolution, but was eclipsed by Darwin's book *On the Origin of Species*. Large jumps in understanding are vanishingly rare, which is why the people who make them become household names. For the most part, we live in the age of scientific refinement, not scientific revolution.

The second pillar of scientific method is testing our new ideas within the constraints of current knowledge. Scientific ideas are called hypotheses, from the Greek word for foundation. Critically, scientific hypotheses need to be falsifiable. While desirable, an experiment that can only give you the answer you want isn't an experiment. Proving something is wrong is as important as proving it is right.

This brings us to the third pillar: most of the time our new ideas will be wrong. As the scientist performing the study this can be hard to accept – so much time and effort goes into research – but it is a necessary part of the process. In fact, it is important – sharing information with the scientific community means that other people won't try the same thing. I won't describe all of my wrong ideas here but trust me, they range from the reasonable through the wacky to the downright stupid. Let me give one example: we had a hypothesis that a vaccine against *S. pneumoniae* might be cross-protective against another bacteria called non-typeable *Haemophilus influenzae*. To test this, Dr Matt

Siggins immunised mice with the vaccine and then infected them. Sadly, there was no protective effect.[1]

In the quest to test ideas we perform experiments. In the biological and medical sciences we often look at some of the population, not all of it. This is driven by practicality. For example, when running a vaccine study you don't immunise the whole country to determine whether your experimental vaccine works. Instead, based on the number of infections you expect to see (the attack rate), you take a small number of people and extrapolate from this. Since we are never studying the whole situation we take samples as a window on the world. We must draw inferences from what we observe through these windows and to do this we use statistical analysis. In simple terms, statistics help us know the probability (or chance) that what we see represents the whole situation.

It might be easier to see this in the context of the research question: are fat men more likely to get COVID-19 than thin men? This may sound like a fickle question but understanding which people are at greater risk helps us understand how to treat the disease and where to target interventions. We first reframe it as a testable hypothesis (phrased as a statement rather than a question), so it becomes: fat men are more likely to get severe COVID-19. We then take a sample of fat men and a sample of thin men and look at the number who get COVID-19. We can either do this *prospectively* – identify the men in advance and follow them to see who gets COVID-19 – or *retrospectively* – look at who got COVID-19 and see how many of them were fat men.

But it is all too tempting to find the answer you want to find – we are human after all. The scientific method is structured to ensure that when we do find something new it is actually correct – this includes blinding, replication and having a control group. Blinding is the separation of the person collecting the data from the person analysing it – so in our fat and thin example, one person might record the weight of people and another would record whether they got infected. Replication ensures you haven't just found a freak situation, like flipping ten heads in a row. If my study in London and your study in New York show the same results, then we can be more confident of their veracity. Finally, control groups – where you do everything to the group except administer the actual treatment. This is most commonly seen in a placebo, where controls are given an identical-looking treatment and the outcome recorded. Admittedly, there can be a 'placebo effect' where just taking a pill makes one feel better – which also needs to be controlled for. Returning to the role of weight in COVID-19, data gathered *did* support the idea that fat men were at greater risk of severe disease than thin men.

Now armed with the toolkit of scientific method, we can set about answering the question: why don't we get infected all of the time? There are two elements to this: the environment and host defence. The environment determines whether you encounter the pathogen in the first place: host defence encompasses how your body protects you from the pathogen if you do encounter it.

ENVIRONMENT

It might seem obvious, but you can't get infected with a pathogen if you never come across it. None of us will ever get smallpox, because of its complete eradication from the globe. There are regional variations in pathogen exposure and these can be determined by a range of factors. For example, you cannot catch a vector-borne pathogen if there are no vectors where you live, so you cannot catch malaria in the UK because the *Anopheles* mosquito is no longer native to the UK. Other pathogens are limited to certain countries, either due to deliberate efforts to eradicate them or because of interventions to reduce their prevalence. Clean water removes the likelihood of exposure to a range of gut (also called enteric) pathogens such as cholera and typhoid.

The weather also has an impact: for reasons we don't completely understand, some pathogens are much more common in the winter than in the summer. For example, influenza virus, which has a predictable seasonal pattern in the UK – between October and March. Peak influenza routinely falls between weeks 48 and 52 (i.e. December) each year because air temperature affects the seasonality of influenza. In warmer temperatures the droplets evaporate quicker, increasing the concentration of chemicals in the drop, like spilled Coca-Cola getting stickier if left. This increase in concentration alters the acidity of the droplet, which in turn affects how the virus gets into our cells. In the cooler winter months, the droplets that spread the virus are more stable. The amount of sunlight also contributes to seasonality. The sun can protect us directly, its UV rays killing viruses, and also indirectly, as more sunlight means more vitamin D, which plays an important role in our immune response. Changes in social behaviour also contribute to seasonality – in winter we are more likely to be inside; especially children, who are a key vector for viral spread. Parents of nursery age children will be painfully aware of this. The winter flu season is only a feature of temperate climates – countries in the tropics don't have a winter flu season, because they don't have winter. There are two rather than four seasons in tropical countries: dry and wet. The patterns of influenza are not as well characterised in these countries, but where measured influenza tends to occur more in the wet season.

For many pathogens, social contacts are required for spread – plagues emerged only when humanity settled down into larger communities. The most likely place to get an infection is at home and the second most likely place is at work. Basically, anywhere you spend a lot of time in close proximity with someone else: in the UK, COVID-19 clusters in homes with larger multigenerational families.

Frequency of exposure is also important. If there is a set risk of getting a disease each time you encounter it, then the more times you are exposed to it the more likely you are to catch it. If you do something often enough, rare events will eventually happen. The risk of infection can be quantified as a function of the number of contacts between an infected person and a susceptible person, the length of time they spend together and the likelihood of transmission each time they make contact.

There is, however, an element of chance to who will get sick. But because we are scientists and it doesn't sound great to say the reason you got infected is down to shitty luck, we dissemble and use the word stochastic. This causes some eyebrow raising but unfortunately biology is messy, even when we know a lot about the virus: in the case of newly emerged pandemics, we know even less.

HOST

Of course, exposure isn't everything. Some people can be repeatedly exposed to a pathogen and never get infected; for example, a group of around 140 sex workers in Nairobi remained uninfected with HIV despite frequent exposure to the virus.[2] Critically, there is a difference between infection and disease. The outcome of what happens once a pathogen gets into the body will depend upon the person infected and is determined by a combination of dose, genes, sex, age, behaviour and underlying conditions.

Much of whether exposure to a pathogen leads to infection and disease will come down to the dose. Going back to the root of the word virus, viruses do behave a bit like a poison: the more you get, the sicker you get. If you stand two metres away from someone infected with SARS-CoV-2 while wearing a mask, you will get a much lower dose than if you stand next to them and they cough in your face.

Fundamentally, the pathogen must enter the body to cause an infection. This most frequently happens at the interfaces between us and our environment: the lungs, the guts, the genito-urinary tract and the skin. Transmission can happen directly through contacts (e.g. kissing) and droplets (e.g. sneezing) or indirectly from a contaminated surface (beware the infected lettuce leaf) or via a vector (mosquitoes, again).

Luckily, we possess a whole arsenal of host defences that stop pathogens getting into our bodies. Understanding these defences and how we can augment them underpins our successes in controlling pathogens, particularly through vaccination.

As a card-carrying immunologist, I would like to claim that the sole reason we don't get infections is because of our immune system. This would be entirely out of self-interest: the more important your subject sounds, the more money you can wring from the government! But the immune system is really the last resort – activating it comes with some risk of damage to your body. Many non-communicable diseases are caused by your immune system going rogue – including allergy, arthritis, asthma, Crohn's, diabetes, Lupus and multiple sclerosis. These are often described as autoimmune conditions, because the body attacks itself. There is a Yin and Yang balance: not enough immunity get infections; too much immunity get arthritis. This is clearly a massive simplification, but the idea that you should boost your immune system with supplements, no matter how much internet influencers insist on it, is as unwise as it is impossible. Even if you could boost your immunity, you are just as likely to boost the damaging aspect as the protective one.

So, before we take the plunge and activate our immune system, we use other lines of defence. The first of these is behavioural. One of our most deep-seated behaviours is disgust. It is so deeply ingrained that you don't even physically need to see a disgusting thing to be disgusted.

Let's try, shall we?

What's the most unpleasant thing you can think of?*

Now imagine eating it. Would you want to? Hopefully not. And in not eating whatever monstrosity the dark places of your mind came up with you have stopped yourself getting infected. We find spit, sick and snot disgusting

* For those of you struggling to think of something disgusting, imagine Kiera (my colleague Katie's dog), who ate another dog's poos and vomited them onto the kitchen floor.

because they are all associated with disease. Not going near them reduces our risk of infection. Some of this is learned behaviour; children get so many colds because they haven't yet learned that eating boogers is wrong.

Our disgust instinct can only go so far to protect us. Sadly for us, not all pathogens have an obvious calling card and they may even pass between people *before* symptoms develop.

PHYSICAL BARRIERS

The next layer of defence is the physical. Our bodies have evolved many barriers to stop pathogen entry. One that you might not immediately think of is the skin. Think back to your youth when you grazed a knee and your parent possibly applied iodine, Dettol, Savlon or something else foul smelling and painful. Often the treatment would feel worse than the original injury, but it did stop the bad things getting into your body.

You can live with an extremely aggressive pathogen on your skin (e.g. MRSA), but the bugs on your skin become a problem when the barrier of the skin is breached – this can be something as minor as a splinter or something considerably more serious, such as a bomb blast injury. Bacteria often hitchhike on foreign objects into places they shouldn't be. One of the most common causes of infections is intubation – inserting a plastic tube into a human, like a catheter. Patients who need catheters to help them pee are at risk of UTI (urinary tract infection). Bacteria love a solid surface so they can form something called a biofilm – examples of this include the black gunk you find around a tile, or the white gunk you get on your teeth. These biofilms grow up catheters and are resistant to drugs, so infection with them is more dangerous – especially as catheterised patients find themselves more vulnerable to infection in the first place.

When we aren't introducing infections into them on Trojan horses, our bodies form a pretty effective barrier. They use a molecular glue called tight junctions, which lock the skin together into an impenetrable layer. Tight junctions are a good example of the rolling evolutionary battle between the infector and the infected, described by the Red Queen hypothesis. We evolved proteins to keep bacteria out but bacteria evolved ways to break these proteins down. Strains of the bacteria *Escherichia coli*, which cause explosive diarrhoea, break the junctions between cells, allowing them to get into the body.*

Protection is more than skin deep: there are lots of barriers that prevent pathogens from entering the body. In fact, the skin has it relatively easy because its sole purpose is as a barrier, stopping stuff getting in or leaking out. Our lungs serve a very different purpose – to exchange gases. We need oxygen from the air to get into our body and carbon dioxide from our body to get back into the air. To enable this, the lungs look like a fine mesh, maximising the exchange of gas per breath. Unfortunately, this provides a large area for viruses and bacteria to land on and cause trouble, which in turn means we need extra protection in the lungs to stop infection. The first of these protective measures comes as a lucky by-product of breathing – longitudinal flow. Pathogens take time to bind to and enter our cells and the movement of air in and out of the lungs can help to dislodge them. This is aided by the cilia; tiny hairs on the surface of the lung cells

* *Escherichia coli* is disappointingly not named after MC Escher.

that sweep things from the lungs up and out. Longitudinal flow also prevents bacterial entry at other points. For example, urinary tract infections are much less common if you wee regularly – you literally piss the bacteria off.

MY CHEMICAL ROMANCE

Beyond the physical and mechanical mechanisms for preventing infection, our bodies are also masters of chemical warfare against the invading hordes. As you'll know if you've ever tasted sick (note the disgust reaction), the stomach contents are highly acidic. This acid bath very effectively kills infectious bacteria in the stomach. Antacids, commonly taken to prevent heartburn, work by reducing the acidity in the stomach, but with a side effect of increased stomach infections. Reducing the acid in the stomach makes it a more hospitable environment for the bacteria.

Another common chemical agent that prevents infection is mucus – or snot. And we don't just produce mucus in our noses – it lines our lungs, our guts and our reproductive tracts. Mucus comprises a complex mixture of sugars, proteins and DNA. The proteins that make it sticky are called mucins. Our bodies make different types of mucus at different sites: airway mucus needs to be thinner and more penetrable than gut mucus because of the difference in function between the lungs and the guts.

If you want to make mucus at home, you'll need 5 g of pig mucin, 4 g of salmon sperm DNA, 5 g of salt, some egg yolk and some milk powder. I suspect your first question might be: where do I get salmon sperm DNA and pig mucin from, as I am not some kind of fish and swine wrangler? To which the answer is the Sigma Catalog, the go-to source for substances chemical and biochemical – with everything for the more discerning customer including Pseudo™ Corpse Scent, strychnine, solid gold and pure cocaine (admittedly you need a licence to order most of the more out there things). Your second question is probably: why would I want to make it at home? I would ask back, what else are you going to do with your time? The scientific reason is that we need to understand how bacteria behave in the actual fluids we find in our body, rather than in the artificial conditions in which we tend to grow them in the lab. In particular, we want to understand what changes when airway mucus changes; if the mucus becomes too thick, you are more susceptible to infection, as seen in cystic fibrosis (CF).

CF is a genetic disease. Most patients with CF have a single mutation in a gene called CFTR (the cystic fibrosis transmembrane conductance regulator). The normally functioning CFTR molecule is a chloride channel which acts as a tube in the cell membrane that selectively lets out chloride ions (the Cl^- in NaCl or common salt). This flow of Cl^- ions leads to water moving from the body into the airways, diluting the mucus. The most common mutation is called F508del (or ΔF508), which in genetic speak means a deletion of the 508th amino acid (the Δ is the Greek letter Delta; genetics shorthand for a deletion). The three letters in the DNA that should encode a phenylalanine amino acid there (the F508) are missing. This affects the protein structure. Some amino acids are more important than others: it's a bit like a Jenga tower – you can take out some bits and nothing happens but remove the key piece and the whole structure comes down. The single ΔF508 deletion – one amino acid out of nearly fifteen hundred – is enough to break the whole CFTR protein. Patients with CF produce much thicker mucus because the faulty pumps fail to dilute it. Sadly, this thicker mucus is a perfect place for bacteria to grow and individuals with CF experience severe bacterial lung infections. Life expectancy for a patient with CF is thirty-seven years and thanks to antibiotics this is three times what it was in the 1950s. There is some hope on the horizon in the form of a drug called Kaftrio, which stabilises the mutant protein (basically fixing the wobbly Jenga block in place). What we learn from this is that people working in pharmaceutical companies can come up with some brilliant drugs but call them bloody stupid things. The other prospect is gene therapy – where the faulty gene can be replaced by a working copy; progress is being made but it is still some way off (see Chapter 15).

Mucus is just one part of a formidable stockpile of antimicrobial chemicals. We also use enzymes to destroy pathogens. Enzymes are proteins that speed up chemical reactions in the body. German physiologist Wilhelm Kühne named enzymes after the Greek word meaning yeast, because yeast accelerated the fermentation of fruit into alcohol. The enzymes that protect us against infection are mostly proteolytic; that is, they chew up other proteins. In an efficient twist, the enzymes that digest our food are also effective at breaking down the components of bacteria in the stomach.

As well as the multifunctional digestive enzymes, we make enzymes that specifically target bacterial biochemicals. One of these enzymes is called lysozyme, which chews up the bacterial walls. It targets a chemical called peptidoglycan, which holds the bacteria together. Lysozyme is present in human breast milk, which reduces the risk of diarrhoea in babies, and in tears, where it prevents conjunctivitis. It is highly abundant in the whites of chicken eggs, which is why they go opaque when you cook them: the proteins all tangle up together when heated. As with tears and milk, hen eggs contain lysozyme to prevent bacterial infection: eggs are stuffed full of nutrition intended for the baby chick, which can easily become a food source for bacteria. Eggs go off if not kept in the fridge because the lysozyme degrades, leaving the egg prone to infection. The abundance of lysozyme in an easily accessible source meant it was one of the earliest proteins to be studied in depth.

As well as active processes that deliberately target and kill bacteria, we use more passive processes in an attempt to starve the bacteria. In order to grow, replicate and survive, pathogens need access to the fundamental elements of life, which are carbon, hydrogen, oxygen, nitrogen, sulphur, phosphorus, potassium, sodium, magnesium, iron, calcium and manganese, with traces of zinc, cobalt, copper and molybdenum.* Most elements are available in or on the human body. This abundance of nutrients puts us at risk of being overwhelmed by bacterial infection: therefore we evolved to restrict their availability to the bacteria. Eugene Weinberg coined the term 'nutritional immunity' in 1975 to describe the mechanism by which our cells stop bacteria getting their hands on our nutrients.[3]

For example, iron is critical to the function of many biological molecules. Humans mainly use it in red blood cells to shuttle oxygen around the body, which is why a shortage of iron in your diet can lead to fatigue (anaemia). Bacteria require iron for the same function – using oxygen to burn sugars. The level of free iron in the blood is very tightly controlled to prevent infection. In the same way

* I've no idea what molybdenum does. I can barely pronounce it.

that insulin controls the level of glucose in the blood, another hormone, hepcidin, controls the amount of iron. When iron levels are unbalanced infections can increase, as seen in a study in Zanzibar. General malnutrition led to anaemia in children and to resolve the anaemia the children received iron and folic acid. Unfortunately, mortality in the group receiving supplements was significantly higher than in the control group. The authors speculated that malaria infection, enabled by the iron, caused these extra deaths.[4]

In addition to individual elements, bacteria need an energy source to grow. Bacteria are very flexible in the energy sources they can use; some even live off plastic bags. However, human pathogens tend to use energy sources available in the human body – particularly sugars. One approach to prevent infection is to stop the bacteria getting the energy they need. Building on work done by Profs Emma Baker and Debbie Baines at St George's in London, we are interested in the effect of elevated lung glucose on susceptibility to bacterial infection. The concentration of glucose in the airway is normally ten to twenty times lower than blood glucose. We believe that the body uses this as a mechanism to inhibit bacterial growth. The level of glucose in the lungs can change, particularly in people with underlying health conditions, such as diabetes. Evidence to support this came from work done by Dr Simren Gill when she was a PhD student in my lab. First Simren showed that diabetic mice were more susceptible to lung infections. She then genetically engineered bacteria, blocking their ability to use glucose. These mutant bacteria did not thrive in the diabetic lung.[5] This is of interest because it offers a way to reduce lung infections in a vulnerable group. Emily Brown in my group works with Professor Matthew Fuchter (a chemist) to design drugs to block this pathway.

Returning to that breath you inhaled at the beginning of the chapter, which hopefully you have now exhaled, you should now feel more confident that it won't lead to you being infected. There are lots of ways in which the body is protected from infection. Curiously, one of the major things that protects us from infection with microorganisms is other microorganisms. We can find a whole community of things living on and in our body. They interact closely with the immune system and are collectively called the microbiome. And we next turn our attention to this fascinating area.

CHAPTER 4

The Microbiome

Timeline: Late May 2020. Epsom (not Barnard Castle). Lockdown continues, for most. Global COVID-19 cases 5,817,385; deaths 362,705.

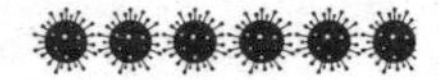

'Humanity was saved by the littlest things which God and His wisdom had put upon this Earth'

H.G. Wells

As the Martians discovered in *The War of the Worlds*, it's a microbial war out there. In the same way that lions compete for prey on the plains of the Serengeti, microorganisms compete for nutrients and space in your body. And believe me, it is a fierce competition. Estimates vary but the numbers of things living on and in our bodies are mind-blowing. Until recently, it was widely stated that your body has ten times more cells of bacteria than it has cells of you. The ratio is probably closer to 1:1, depending on whether you measure the bacterial number before or after someone visits the toilet. A recent recalculation suggests that we are made of about 3×10^{13} (3 with 13 zeros) human cells, most of which are blood cells, and we carry 3.8×10^{13} bacteria cells, most of which live in the guts. But 38,000,000,000,000 is still a lot of bacteria – about 0.2 kg in fact.[1] For the bacteria you are basically a walking buffet.

The last twenty years have witnessed a revolution in our understanding of the bacteria associated with our bodies, which has been driven by incrementally improving technology. This mirrors a larger revolution in the way in which biological questions are addressed, with a huge increase in the availability of data. Biology isn't unique in this, as 'Big Data' increasingly dominates all fields, from government to healthcare to shopping. A combination of computing power, computer storage and data generation all underpin Big Data.

We use the catch-all 'omics to describe Big Data techniques employed in biology. The word 'omics comes from the suffix -ome, meaning all of the constituents of a set considered collectively. One of the first uses of -ome was in genome, which is the genetic material of an organism. But we can also study the transcriptome, metabolome, epigenome, proteome, lipidome, glycome and connectome. There are more niche studies, including those of garden ornaments and extra-terrestrials: the gn-ome and the phone-ome respectively.

These new 'omic techniques litter the pages of the glossy science journals and are proving incredibly useful. Yet they do not lack their critics. Sydney Brenner, who won a Nobel Prize for studying the nematode worm (*Caenorhabditis elegans*), described them as 'low input, high throughput, no output science', by which he means that generating data for the sake of generating data does not necessarily help us to answer biological questions. It can be challenging to analyse and interpret the Big Data in a meaningful way. As described, we use statistical tests to understand whether what we have observed truly reflects the real world. But the more statistical tests you perform the more likely you will get a positive result. When you analyse a set of data with a million points, comparing everything against everything else can lead to some misleading results. Therefore, along with the new tools for generating the data, we need new tools to analyse and interpret the data. We also need new tools to present the data so other people, both scientists and policymakers, can understand our results. A bar chart with two groups is easy to understand and present, but a network of interactions less so.

A particular sin associated with 'omics data is confusing causation with correlation. Correlation (literally co-relation) is a relationship between two things. For example, ice cream sales correlate with temperature: the hotter it gets, the more people want ice cream. However, just because two things correlate, it doesn't mean that one causes the other, or that they are linked at all. Numerous things correlate nonsensically: for any year between 1999 and 2009, the number of films Nicolas Cage appeared in correlates with the number of people who drowned in swimming pools; the number of civil engineering PhDs awarded correlates with the consumption of mozzarella; and the number of letters in the winning word of the national spelling championships correlates with the number of people killed by venomous spiders. These are obviously nonsense, though statistically true. The problem with biological data is when plausible things correlate but the relationship is not causative, leading to the catechism 'correlation is not the same as causation'. Unfortunately, this sin of falsely inferring causation occurs frequently in the -ome described here – the microbiome.

Some scientists use the microbiome solely to refer to the genes of the microorganisms that live in a specific evolutionary niche, rather than the microorganisms themselves. A more specific name for all of the microorganisms that live in an environment is the microbiota. But most people refer to these microbial communities as the microbiome and that's the way I will use the word. Researchers study the microbiome of a huge range of things – shower heads, plant roots, teeth, beards. In the context of infection, we are interested mostly in the human microbiome – the bacteria that live on and in our bodies.

So why the interest in the microbiome in the context of infections? The answer is twofold. Firstly, many of the bacteria that can infect us can also live in peaceful harmony with us. They only become dangerous when the system is perturbed. Secondly, the microbiome forms part of our defence against bacteria. For bacteria, the human body represents a treasure trove of things they need to eat, grow and survive, particularly in the guts. Because all living things use the same basic building blocks (sugars, fats, proteins, micronutrients, etc.), the contents of your stomach represent a great source of food for bacteria. Which explains why we find so many bacteria in our poo. These bacteria will carry on living after you die. If Henry VIII did really explode in his coffin, it would have been because the bacteria in his guts had carried on fermenting and without royal flatulence the gases had built up to catastrophic levels.

While the guts contain a rich source of nutrients, they can only support a finite amount of bacteria. This leads to competition between microbes. If you are Mr *E. coli*, you want to take all of these goodies for yourself and not share any with anyone else. As a consequence, microorganisms evolved tools to help them kill other microorganisms. Antibiotics are an example of this; yeast produces penicillin to kill bacteria. Bacteria make proteins called bacteriocins, which stop the growth of other closely related bacteria. Which makes sense – the species most likely to take the resources you want is the one most similar to yourself. It is the microbial equivalent of vloggers trolling each other to get more hits.* This internecine warfare works out well for us – the bacteria keep each other in check.

Different nutrients are available on different surfaces, creating a micro-environment which will support some bacteria better than others. The micro-environment in your small intestine differs greatly from the micro-environment in your lungs and so we observe differences in the bacteria that live there. Studying these bacteria is a form of ecology – as the grass feeds the rabbit that then feeds the fox, so we can investigate the interplay of bacteria and other microorganisms in human samples.

One of the striking things is the incredible diversity of bacteria that make up our microbiota. We find diversity between individuals and even closely related individuals – if you measure the microbiome from two identical twins, it will be different. We also see differences within individuals. If you haven't cleaned your teeth yet today, scrape your finger on them – that goopy biofilm referred to as plaque in toothpaste commercials consists of a web of bacteria. When researchers sequenced the bacteria in tooth plaque, they found that each tooth possesses its own unique bacterial signature.

* And because I am down with the kids, I know that these are called diss tracks – cue eye rolls from my son.

So, where does this bacterial diversity come from? Before birth, we are more or less sterile; an empty canvas to be colonised by microbes. I say more or less sterile, because some people have described the presence of bacteria in the womb. This is difficult to prove conclusively because there are so few bacteria present. In order to find rare things, you need highly sensitive techniques. Using highly sensitive techniques to detect things can generate misleading results because bacteria can be found everywhere. So in the process of measuring the bacteria in your sample you can accidentally measure bacteria in the swabs you used to collect the sample and in the scientific kits you use to process the sample – sometimes referred to as the kit-ome.* Either way, at birth the amount of bacteria on or in us is next to zero.

The nature of your birth profoundly shapes the bacteria that live with you for the rest of your life. Those of you who have given birth or been near it when it happens will know it is a pretty messy process. The baby hole being pretty close to the poo hole means that as the baby passes through the birth canal it acquires the bacteria from both the vagina and the rectum. Children who come out of the sunroof (born by C-section) have a less diverse microbiome at birth, but this recovers pretty quickly as they crawl through the microbe-laden world. This may sound gross, but bacterial diversity at birth predicts a whole range of health outcomes in later life.

Newborns also acquire bacteria from the mother's skin when breastfeeding. Curiously, breast milk contains special sugars that only bacteria can digest, indicating the extremely close relationship between man and his microbes; woman

* Niche 'omics joke.

too, but there was no convenient catch-all word meaning bacteria starting with 'w'.

While early life events mould the microbiome, our fellow bacterial travellers do change throughout our lives. A whole range of factors alters the human microbiome, including smoking, coffee consumption and bread preference.[2] Air travel, particularly when crossing time zones, can also alter it. This is because, like you, your gut bacteria have a body clock and therefore can get jet-lagged. To demonstrate this Eran Elinav, from the Weizmann Institute in Israel, sent two of his PhD students on a plane and sampled their microbiome before and after flying to show how it had changed.[3]

Less exotically, my group looked at the impact of infection on the gut microbiome (mostly because we don't have the budget to fly students around the world just so we can sample their poo). Dr Helen Groves, as a PhD student in my lab, investigated how lung infection altered the gut microbiome. She did this in collaboration with Professor Miriam Moffatt and Dr Mike Cox and his magnificent beard. Helen observed that following an influenza infection in the nose, the bacteria in the guts of mice change. We wanted to understand why, as it could impact how animals recover from infection. Helen demonstrated that loss of appetite following infection caused the change in gut bacteria.[4] We now want to understand what these changes mean. One entirely speculative hypothesis is that by changing the gut microbiome you recover quicker, though we have no evidence to support this, yet. But Zee Wang in my group, who is just starting her PhD, is investigating why infection makes us stop eating; she is trying to get to the heart of the feed a cold / starve a fever idea.

We rely upon our microbiome for a whole range of functions. Humans and their bacteria form a mutualistic relationship – both species gain. You may be more used to mutualistic relationships from big budget nature documentaries. The following examples will work better if you use your best David Attenborough voice: take Ritter's sea anemone and the *Finding Nemo* fish* – the fish protects the anemone from butterflyfish while the anemone stings potential predators of the fish; or the bee and the flower – the bee pollinates the flower while collecting nectar for itself; or the oxpecker bird and the zebra – the bird eats the ticks from the skin of the zebra, getting food for itself and reducing the parasite burden for the zebra.

The clearest mutual benefit between humans and bacteria is in digesting the food we eat: without bacteria, much of the fibre in our food would remain indigestible. Bacteria also play a role in the uptake of fats from the guts via chemicals called bile. There are two steps to bile production: human cells make the initial precursor molecule, but we need gut bacteria to break it down to produce the active chemical that soaks up fats. As a side effect, bile also makes your poo brown. As well as helping digestion the gut microbiome influences appetite, interacting with the hormones that give you feelings of fullness after a meal. Some strains of bacteria make you feel fuller after eating than others; transferring bacteria from the guts of a fat mouse to a thin mouse leads to overeating in the previously skinny mouse.

The microbiome also trains the immune system. Mice bred with no microbiome (called germ-free) have very screwy immune systems and are more prone to infection than mice

* Clownfish, if you insist.

with an intact microbiome. These germ-free mice give us insight into the role of bacteria in the development of asthma, first described as the 'hygiene hypothesis' by Professor David Strachan at St George's Hospital, London in 1989. He hypothesised that as the rates of childhood infection decreased, rates of allergy increased. Evidence to support this idea came from a study by Professor Erika von Mutius on the relative rates of asthma in East and West Germany after the Berlin Wall came down. This was a useful natural experiment because the two populations were genetically similar and therefore environmental factors could be investigated. She found that despite there being much higher levels of air pollution in the former East Germany, there were only half as many allergies. Investigating further, von Mutius linked allergy to contact with farm animals in early life. She suggested that exposure to a diverse range of bacteria in early life shaped the immune system later on.

Further evidence comes from comparing rates of allergy in two American farming communities – the Hutterites and the Amish. Both of these communities are relatively isolated genetically (a polite way of saying they have high levels of inbreeding). Hutterites and Amish have relatively similar genetic backgrounds (they both came from Europe in the sixteenth century), but while the Amish famously eschew twenty-first-century technology (and twentieth-century technology and most of the nineteenth-century technology to boot), the Hutterites use modern farming practices. As a consequence, the Amish and their children live much closer to their farm animals and in line with the 'hygiene hypothesis' there is a proportionally lower rate of asthma in Amish kids, which possibly makes up for not having iPads, washing machines or electricity.

While we do not fully understand the mechanism, evidence suggests that early exposure to bacteria can be beneficial, in part because the first foreign microorganisms our bodies encounter act as a training set. Because of the observed links between the microbiome and health, much has been made of improving microbiome diversity and avoiding 'dysbiosis', whatever that actually means. The simplest approach is to increase dietary diversity by eating foods that bacteria like; for example, high fibre food. For those who don't like All-Bran, a whole industry has been developed to nurture your bugs. For example, prebiotics: fibrous compounds that support bacteria growth in the guts. They may help supplement a diet lacking in fibre, but beware – providing a kick-start to your gut bacteria will also kick-start their waste production, with noxious effect. Alternatively, probiotics can be used, which are foods that contain live bacteria; like yoghurt or any of those disgusting things beginning with 'k' (kefir, kimchi or kombucha). While this sounds like a compelling idea, there is little evidence that they have any impact; potentially because the competition for resources in the guts is already so fierce that any incoming bacteria can find no space to establish themselves. They certainly shouldn't be seen as a replacement for a balanced diet.

There is probably a space for encouraging some exposure to commensal bacteria in early life and an over-reliance on 'antimicrobial' soap may also be problematic.* As with all things, balance matters. I need to stress that the hygiene part of the hypothesis does not refer to personal cleanliness. Not washing your hands before eating will not stop your

* All soap is antibacterial – it's kind of the point.

children getting asthma, but it is going to give them diarrhoea. Poor personal hygiene contributes to the spread of all kinds of infections. The simple measure of handwashing during the COVID-19 pandemic slowed not only the transmission of SARS-CoV-2, but a whole slew of other viruses too.

All told, microorganisms (good and bad) surround us and we need some way to keep them out. At the end of *The War of the Worlds* the plans that the Martians had slowly but surely made against us fell apart because they didn't have an immune system. And so we now turn to this key arbiter of health and disease.

CHAPTER 5

Immunology

Timeline: Mid-June 2020. In London in my lab which has reopened. Global COVID-19 cases 7,823,289; deaths 431,541.

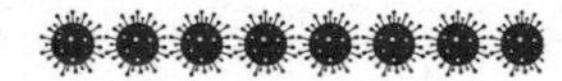

'Fight the good fight.'

Rev. John Samuel Bewley Monsell

When I was seventeen, I wanted to impress the interviewers at Cambridge University. I thought the best way to do this would be to read random science books to give me an air of sophistication. The first one I picked up covered the immune system and something clicked because twenty-five years later I am still working on the same subject. Sadly, try as I might, I can't remember the book's name and give credit to the author for sending me down the path I am on, but thank you anyway (conversely, at times when academia gets on top of me it is probably best that the author remains anonymous so I can't blame them for my life choices).

As mentioned in the previous chapters, pathogens must overcome multiple barriers to get into a human body, but occasionally they get through. At this point we need something to contain, control and ideally clear the infection – our immune system.

Immunity is the ability to tell self from non-self at the microbiological level. It is hard-wired into the genes of every living organism, from the biggest to the smallest. It may sound surprising, but even bacteria can get infected. The family of viruses which infect bacteria are called the bacteriophages. Credit for the discovery of bacteriophages lies with Félix d'Hérelle, a French-Canadian microbiologist with a rather striking beard/moustache combo that a modern-day hipster would be jealous of, proving that microbiology was cool before cool was even a thing.* He taught himself science and led a remarkably colourful life. Amongst other things, he lost all his money in a failed chocolate factory, worked in Guatemala, Mexico, France, India, the USA and Egypt (not trivial before international flight), had a brief run-in with the secret police in Soviet Russia when his mentor fell in love with the same woman as Beria (the notorious head of the NKVD) and was put under house arrest by the Wehrmacht in the Second World War. In 1917, d'Hérelle observed that you could kill bacteria with an extract from chicken poo and that the extract could still kill the bacteria even after filtration through the same kind of superfine filter that Ivanovsky and Beijerinck had used to discover plant viruses. He called the filtered extract **bacteriophage** – combining bacteria with the Greek word *phage*, meaning to eat. His work was not without controversy. Along the way d'Hérelle antagonised a Nobel Prize winner called Jules Bordet, who had discovered a different (but not

* Of all the scientists I learned about while researching this book, Félix d'Hérelle was undoubtedly my favourite.

necessarily mutually exclusive) way of killing bacteria called complement. This precipitated a ten-year-long feud. Bordet championed a British bacteriologist called Fred Twort as the discoverer of the bacteriophage. For what it's worth, Twort probably did first discover the phage but didn't pursue his findings, being distracted by the minor inconvenience of the First World War. The feud culminated in d'Hérelle challenging Bordet to a scientific duel in the academic press. Science in the early twentieth century was more interesting than now – a number of scientists have enraged me, but I have never thought to resort to duelling with pipettes!

The important point of that diversion into bacteriophages and the remarkable life of Félix d'Hérelle is that bacteria can get infected by viruses. The bacteriophages are the most abundant type of virus on earth, which makes sense if you consider how many bacteria live on a single person, let alone globally. To stop viral infections, bacteria evolved a proto-immune system that recognises when they are infected and destroys the genetic material of the virus. One example is called the restriction/modification system. This system uses proteins called restriction enzymes, which cut DNA in precise places. Each enzyme targets one specific gene sequence: for example, the enzyme *Eco*RI cuts DNA at a palindromic sequence: CAATTG.* Bacteria protects itself from auto-destruction by modifying its own DNA so that

* Which is not a palindrome unless you consider the way that DNA pairs:

CAATTG
GTTAAC

The top strand is reflected in the bottom strand.

the restriction enzyme can't cut it. Scientists have adapted the tools bacteria use to destroy viruses: restriction enzymes have been the backbone of DNA engineering for the last thirty years. Yet it should be noted that these enzymes are sensitive to temperature and if you leave a box full of them out of the freezer on the lab bench overnight you will most likely get a right bollocking from your supervisor (according to a 'friend' of mine).

Another system with which bacteria protect themselves is called CRISPR-Cas9. CRISPR refers to a region in bacterial DNA and it stands for *clustered regularly interspaced short palindromic repeats*, which is why everyone calls it CRISPR (I say it crispER, but other people put the emphasis on the R as in crispR). Cas9 is a protein which, like the restriction enzymes, cuts DNA at a very specific site. It differs from the restriction enzymes because the sequence it recognises can change. Cas9 uses a piece of RNA called a guide to identify where it should cut. These RNA guides act as a form of immune memory: bacteria can bank viral RNA so the virus will be recognised by future generations.

The good news for fans of scientific intrigue is that the discovery of CRISPR-Cas9 doesn't lack controversy. At least four groups claim precedence and, more lucratively, the intellectual property rights: sadly, no duels yet. In 2020, the Nobel Prize committee decided that the credit belongs to Emmanuelle Charpentier and Jennifer A. Doudna. Charpentier worked on *Streptococcus pyogenes* at the Max Planck Unit for the Science of Pathogens in Berlin, Germany. In 2011, she identified the guide RNA used by the bacterial system and then, working with Doudna, demonstrated that this could be used to cut DNA precisely. They also demonstrated that you could reprogramme the guide molecule so that you could specifically cut any DNA sequence. The CRISPR-Cas9 system can also be adapted for gene engineering. You can instruct the Cas9 protein to cut the DNA wherever you want. When Dr David Busse, as a student in my research group, wanted to know how a gene called IFI44 worked, he used CRISPR-Cas9 to cut the gene out of some cells.[1]

Of course, bacteria are not the only organisms with immune defences against infection. Immunity exists in all living things. Scientists who study the immune system are called immunologists. Nobel Laureate Ilya (also called Élie) Mechnikov coined the word immunology from the Latin word *immunis*, meaning exempt. Mechnikov was the first person to investigate the role of white blood cells in preventing infection. He worked on starfish, history doesn't record why, and showed that when you stick a thorn into starfish larvae cells gather at the site of the invasion. Mechnikov called these cells phagocytes – from the same Greek word *phage* used in bacteriophage, referring to the cells' ability to eat other things. The suffix -cyte also derives from the Greek and means cell; in our blood we have red cells which carry oxygen (erythrocytes) and white cells which form the immune system (leukocytes).

As creatures get more complicated, their immune systems get more complicated. Vertebrates have the most sophisticated immune systems of all: the more complex the vertebrate, the more advanced its immunity. Mammals have a more complicated immune system than birds, who have a more complicated immune system than frogs, who have a more complicated immune system than fish.

Immunology can be regarded as a tricky topic. At least in my experience, medical students roll their eyes when I come to teach it to them. I put this down to the subject matter itself rather than my *presentation* of that subject matter. Though, just possibly, my presentation could be improved beyond slides that come in a somewhat random order, PowerPoint colour schemes from the 1990s (yellow on blue a favourite) and the inevitable acceleration through the lecture as I realise I've spent too long on interesting anecdotes about fascinating French-Canadian phage-lovers and not on the learning objectives that I made up to appease the course director. Luckily the students pull no punches. 'It's like he doesn't know what slide's coming next', was pretty close to the mark. Moving my lectures online in 2020 ironed out some of the kinks as I really had to focus on the sequence of the lecture, but I did discover that I start nearly every sentence with the word 'so'.

So, immunology is not inherently tricky, but immunologists love categorising things into smaller and smaller subtypes with obtuse acronyms. The acronyms may be historical, but I sometimes suspect immunologists make it deliberately complicated to restrict membership to the club: our shibboleth.*

THE HUMAN IMMUNE RESPONSE

The immune system comprises white blood cells (leukocytes) that patrol the body looking for pathogens. I am going to focus on some of the more critical ones – neutrophils, macrophages, T cells and B cells (see figure 4).

* Or is that sibboleth?

Innate immune cells

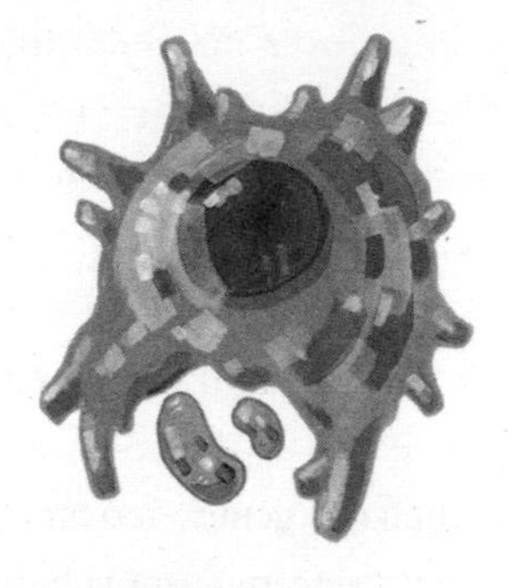

Macrophage: Eats stuff

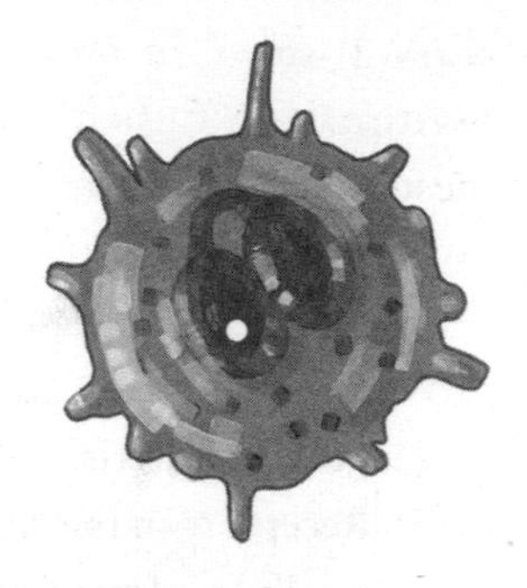

Neutrophil: Kills stuff

Adaptive immune cells

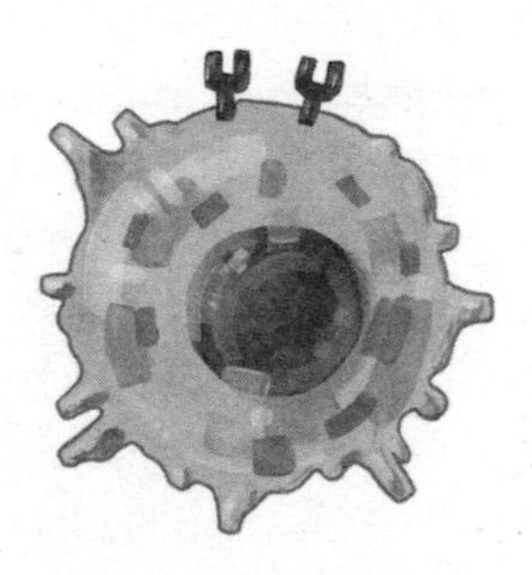

T cell: Orchestrates immune response/ Kills infected cells

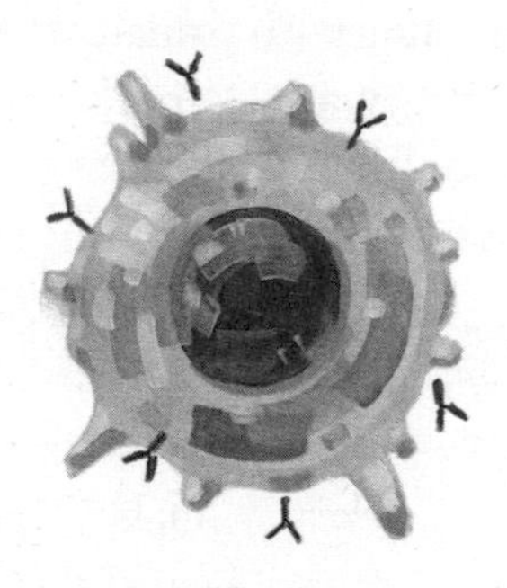

B cell: Makes antibody

Figure 4 Components of the immune system: White blood cells are divided into two classes – innate, which recognise core features of pathogens; and adaptive, which recognise specific features of individual pathogens and have memory.

The easiest way to understand the immune response is in the context of an infection (see figure 5). Because SARS-CoV-2 was very much on everyone's mind at the time of writing, I will use it as an example. The sequence of events described here by and large applies to most infections.

1. The immune response to SARS-CoV-2 initiates when the virus enters the cells by its spike protein and releases its genetic material.
2. Receptors in the cell recognise the viral genes, activating a cellular alarm system. This alarm system dampens the replication of the virus.
3. The alarm system also recruits white blood cells to the site of infection, particularly neutrophils and macrophages.

This initial phase of the immune response is called innate immunity. The innate immune cells can contain the infection, but they are often insufficient to completely clear it. In order to do that, we need a second layer of response described as adaptive immunity, which is composed of T cells and B cells.

4. The T cells identify virus-infected cells and kill them.
5. The B cells produce antibodies specific to the coronavirus spike protein, blocking any new virus from entering other cells and thereby preventing new rounds of infection.

Having killed the virus, the adaptive immune system generates a memory of that particular strain so that you cannot get reinfected in the future. Following that loose framework, I will go into the individual parts in depth. But I'd recommend marking the next page that contains Figure 5, for reference.

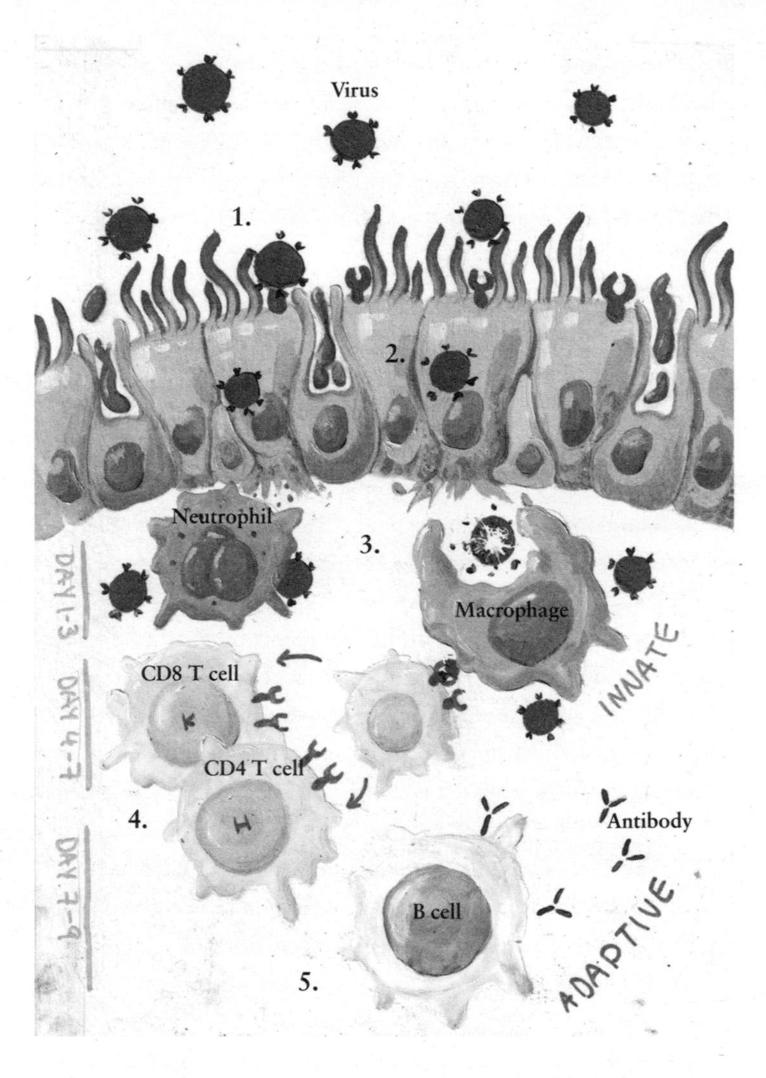

Figure 5 An immune response: Steps correspond to the text. 1. Virus enters cell. 2. Cell alarm system detects virus. 3. Recruitment of innate immune cells (macrophages and neutrophils). Adaptive immune response initiated. 4. CD4 T cells orchestrate response and CD8 T cells kill infected cells. 5. B cells release antibody.

INNATE IMMUNITY – ANCIENT ENEMIES, ANCIENT DEFENCES

Now to look at the parts of the immune response in detail, starting with the innate immune system. Innate immunity is the primal first line of defence. It is found in all multicellular organisms and recognises core immutable characteristics of pathogens, triggering a cellular alarm system. The idea of common characteristics may sound at odds with my description of the immense variety of pathogens in Chapter 2, yet at a biochemical level there are features common to all bacteria or all viruses not present in human cells that helps us to recognise them. It is the equivalent of how you can recognise a human as a person rather than a teacup, even though individual people (and teacups) can be very different.

These unique biochemical features are described as pathogen associated molecular patterns (PAMPs for short). They are often structural components unique to pathogens, not human cells; for example, bacteria have a cell wall while human cells only have a cell membrane. Charlie Janeway, an immunologist at Yale often called the father of innate immunity, first proposed the idea of PAMPs, i.e. uniquely foreign signatures of infection. Janeway also achieved the ultimate accolade – he wrote a textbook so universally used it became eponymous: *Janeway's Immunobiology.*[2*] Polly Matzinger slightly modified the idea to be more inclusive and cover danger rather than pathogens alone. Matzinger famously published a scientific paper with her dog,[3] Galadriel Mirkwood, as a way to get around being the only

* See also *Gray's Anatomy* (which the TV series is named after) and Alberts's *The Cell.*

author on the paper. This led to the editor banning her from the journal for life; a shame because hidden Easter eggs in scientific papers are a joy. In a now revised study about sequencing the baboon gut microbiome, the authors had sneaked the then president's face onto the monkey's faeces.[4] Sadly they revised the image, but if you search for 'paper with Donald Trump's face on a poo' you can still find it.

Cells recognise these molecular danger signals using proteins called receptors. Receptors play an incredibly important role in all aspects of the body's response to the environment – including light receptors, smell receptors, taste receptors, pain receptors, etc. Receptors also facilitate signalling between cells. The classic analogy is a lock and key, but that doesn't quite capture it: the recognition process is certainly specific like a lock, but it is closer to a child's toy where the cow fits into the cow-shaped hole, leading to a moo.

The first pathogen-associated danger signal to be characterised was a molecule that makes up the cell walls of some bacteria. Unlike human cells, bacteria are surrounded by an additional wall (the basis of the Gram stain described in Chapter 2) made of a molecule called lipopolysaccharide (LPS). The receptor that recognises LPS is called TLR4 (toll-like receptor 4). Jules Hoffmann and his team first identified TLR4 as part of the fruit fly immune system. For the work on TLR, Hoffmann would eventually share the Nobel Prize in 2011 with Bruce Beutler, who demonstrated its role in mammalian immune systems. This wasn't the first time the humble fruit fly (*Drosophila melanogaster*) had earned science's biggest prize. Work on *Drosophila* contributed to six Nobel Prizes, covering a range of bodily functions including how our body tells the time, how genes are inherited, how X-rays cause mutation, how embryos develop and even how our sense of smell works.

One of the really fascinating things about the innate immune system is the similarity between different species (highly conserved in evolution speak). For example, while we only share 60% of our DNA with fruit flies, we use *exactly* the same gene (TLR4) to recognise the same bacterial component (LPS). The innate immune system remains unchanged between flies and man because there are some things that pathogens cannot change, including the materials they use to encode their genes or build their walls. Since pathogens have been infecting other cells since the beginning of life on earth, the same fundamental alarm system can be used.

Tripping the cell's alarm system triggers a cascade of responses that serves two purposes – to protect the infected cell and to alert the rest of the body. Much of the local protection is coordinated by an antiviral factor initially described by Alick Isaacs and Jean Lindenmann in Mill Hill, London – which shows that some good things can come out of North London. While investigating influenza virus replication Isaacs and Lindenmann observed that pre-treating chicken cells with dead virus prevented subsequent infection with live virus. They showed that this effect was transferable: when they took the liquid in which the cells were grown and transferred it to another flask, the new cells remained protected against infection. They called this transferable factor interferon. It is cool stuff. It even makes its way into science fiction – the heroes in Michael Crichton's *The Andromeda Strain* use it to prevent themselves from dying of a deadly space-borne virus.[5]

Interferons are critical in preventing viral infections and viruses invest a lot of effort in trying to overcome them.

Remember how cow RSV can't infect human cells and human RSV can't infect cow cells unless we perform some molecular jiggery-pokery and switch over a viral protein called NS1 (see Chapter 2)? NS1 suppresses the host interferon response, but only that of the relevant species – cow NS1 suppresses cow interferon and human NSI human interferon. During the COVID-19 pandemic my colleague Dr Vanessa Sancho-Shimizu, working as part of a large consortium coordinated by Dr Jean-Laurent Casanova, showed that 3.5% of the patients with severe disease had defects in the interferon genes.[6]

The incredible importance of interferon leads to a lot of redundancy in the system, so if one component doesn't work properly another one can replace it. The main members of the family include interferon alpha (IFN-α), interferon beta (IFN-β), interferon gamma (IFN-γ) and interferon lambda (IFN-λ). Other fringe members of the interferon family exist, including interferon omega (IFN-ω): sadly, I failed to persuade my students it was called interferon Wu because ω looks like the symbol used by the hip-hop collective the Wu-Tang Clan.

Interferon works locally and systemically. Local interferon production indicates that a cell in the immediate vicinity is infected and flags a risk to neighbouring cells. The cells in close proximity to the infected one enter an interferon-induced antiviral state. Remember that viruses need to hijack the cell machinery to make copies of themselves – to prevent the virus from replicating, interferon instructs the surrounding cells to take a scorched earth approach, burning all the resources that the virus needs. In addition to shutting down the replication machinery in infected cells, interferon boosts the cells' defences to prevent infection in the first place.

As well as protecting cells in the local area, interferon helps to recruit cells to the area where the infection occurs. While most cells are static, e.g. your eye cells always stay in your eye, the cells of your immune system can move around the body in a targeted way. If you get a splinter in your finger, cells will move there – the same as Mechnikov and his punctured starfish. It's hard to visualise this on paper, but if you briefly head over to YouTube and type 'neutrophil chasing bacteria' you will get a lovely video of a cell chasing its prey. This process of cells moving around the body towards a specific place is called chemotaxis. It is not dissimilar to how sunflowers turn towards the sun. A stimulus attracts the cells and they move towards it: in the case of sunflowers the stimulus is light (phototaxis), but for immune cells it is chemicals called chemokines (chemotaxis). Humans encode forty-seven different chemokines, prompting the fun family game, 'I'm a chemokine – what chemokine am I?', in which you guess the protein being described from snippets of biochemistry. The chemokines are very specific: they only recruit cells which display the cognate receptor on their surfaces. I like to imagine them as address labels directing cells around the body.

The process of cell recruitment to the site of infection or damage is called inflammation and it has four characteristics – heat, pain, swelling and redness. You'll recognise these signs if you have ever had a cut, an infection or a splinter. At the time they are annoying, but they stop the infection spreading further. The swelling and redness result from the dilation of blood vessels, which allows cells to move more freely; the heat comes from the body cooking the bacteria *in situ*; and the pain is a side effect of the local swelling and release of chemicals. Pain, though painful, is a helpful reminder to protect that specific area from further damage. A Roman scholar, Celsus, first characterised and recorded these symptoms of inflammation in the first century CE. As with all things immunological, there is good inflammation – stopping infection – but also bad inflammation, which explains why anti-inflammatory drugs such as aspirin, paracetamol or ibuprofen, which blunt the production of inflammatory signals, can be helpful.

GREEDY CELLS MAKE GREEN GLOOP

The first responders can be grouped together as innate effector cells; a fancy way of saying they belong to the innate immune system and do something on arrival. You will know them better as pus. Most of the white or green gloop that comprises pus is dead cells that have valiantly rocked up and died. Foremost amongst these is the neutrophil. Neutrophils are one of the easiest cells to recognise. In fact, even I can spot them and I am truly terrible at histology (the study of cells). However, you will need access to a microscope and a blue dye called haematoxylin, which you can either buy or extract from the logwood tree (*Haematoxylum campechianum*), which is native to Mexico. I find neutrophils so easy to spot because of their distinctively shaped nucleus, which has three to five lobes connected by thin filaments, like beads on a string. As with many things in the body, form follows function. The distended shape means that neutrophils can squeeze through tiny gaps in the walls of blood vessels to get to the pathogen. Once they arrive, neutrophils bring the big guns to fight infections, particularly bacterial infections. This includes vomiting their DNA to make a net that traps the bacteria, the release of poisonous chemicals and eating (or phagocytosing) the bacteria.

Neutrophils aren't the only immune cells capable of eating bacteria. The other great gobbler is the macrophage; if you were paying attention a couple of pages ago you will remember that Uncle Elie Mechnikov first identified them in his starfish. Macrophages are found all over the body: they reside in the lungs, the blood, the brain, the liver, the bones. The main role of macrophages is to remove things by consuming them, including bacteria, broken cells and environmental material – they are the Pac-Man of the immune system. Macrophages collected from human lungs are soot black because they spend their lives desperately trying to remove all of the car exhaust fumes and other crud we breathe in.

There are other cells involved in the innate immune response and I will no doubt be slated by aficionados for leaving them out. But for now we need to move to the other part of the immune response: adaptive immunity.

ADAPTIVE IMMUNITY – CHANGING ENEMIES, CHANGING DEFENCES

Adaptive immunity gives us a tailored response to pathogens and, critically, memories of previous infections. The adaptive immune response is unique to each individual and can be seen as a history of every pathogen you've ever met.

Two cell families make up the adaptive immune system: T cells and B cells. They display an incredible specificity in their response: each T or B cell recognises one pathogen and one pathogen only. The specificity occurs at a molecular level: adaptive cells recognise a single protein made by a single pathogen, for example the SARS-CoV-2 spike protein. The recognition occurs at an extremely fine granularity: a specific T cell recognises a 10–12 amino acid sequence within a pathogenic protein. We call the protein that the cells recognise the antigen and the specific region it recognises the epitope. This will be important when we come to think about vaccines later; SARS spike was the antigen used in most of the COVID-19 vaccines. T and B cells allow us to react adaptively to pathogens, only making a response to the organisms we see and not those that we don't.

It's about at this point in my lectures when my students' eyes begin to roll into the back of their heads as the information overwhelms them. I try to break things up with the occasional dad joke; for instance, telling them that the B in B cells doesn't stand for boring, it stands for brilliant. This is admittedly a weak joke but the response is usually pretty positive, which goes to show that students aren't as sophisticated as they like to think. In my lectures the B cell zinger is accompanied by some pretty shoddy PowerPoint animation that adds to the overall effect, which may explain why you are not laughing now.

Anyway, back to the immunology. I need to cover three crucial topics to set the stage for the rest of the book: HLA, T cells and antibody.

HLA: SCENTS AND SENSITIVITY

HLA stands for human leukocyte antigen and understanding it is a massive topic: the HLA region of the genome contains two hundred genes, or nearly 1% of the whole human genome. The major product of the HLA region is a protein cluster called the major histocompatibility complex or MHC (I did warn you immunologists love acronyms). The HLA gene region is incredibly variable (polymorphic); there are 863,288 possible combinations of the three most important genes in the cluster (with 59 HLA-A types, 118 HLA-B types and 124 HLA-DR types). The combination of HLA genes you inherit from your parents shapes your response to health and disease. Curiously, it may also determine sexual attraction (I would recommend reading Dan Davis's *The Compatibility Gene*[7] after this one if your appetite is whetted). The MHC proteins encoded by the HLA genes work as a quality control system for the cell. Each time the cell makes a protein it takes a bit of that protein and displays it on the MHC complex on the surface of the cell. Following viral invasion, the cell will make and display viral proteins on its cell surface.

This ability to flag danger to the outside world means HLA genes shape our susceptibility to pathogens, because different HLA types display different parts of the pathogens. For example, individuals with HLA type B57 are much less likely to progress to AIDS after HIV infection. The mechanism underpinning this is a good example of the HLA–pathogen interactions and a nice demonstration of how viruses evolve under pressure from the immune system.

The MHC encoded by the HLA-B57 gene presents a specific region of the HIV Gag protein required for the successful assembly and release of baby HIV viruses from infected cells.[8] To escape the immune system, the virus mutates the part of the Gag gene presented by the MHC. This mutation has mixed results for the virus – the immune system no longer kills it, which is a plus, but the mutation reduces the ability of Gag to make baby viruses, which is a minus. The net result is a slower progression to AIDS in people with HLA-B57. Intriguingly, this viral mutation is not permanent. The spread of HIV can be tracked between different people at a genetic level. When the virus spreads from a person who is HLA-B57 to someone who is not, it rapidly reverts to the original form of the Gag protein. This shows that the immune system puts an evolutionary pressure on the virus, forcing it down a route it doesn't want to go, but following the removal of that pressure it can revert to its original form.

EVERYTHING STOPS FOR T

HLA is an important part of how the adaptive immune system recognises infection. But seeing the infection is only half the battle: it needs to be controlled by something and this is where T cells come in. As every parent has a favourite child and every PhD supervisor has a favourite student, every immunologist has a favourite cell type. The difference is that parents and supervisors pretend they don't. T cells are my favourite. The T sadly does not stand for tremendous; it stands for thymus, where they develop. The thymus is an organ found just above the lungs, whose sole purpose is producing T cells. T cells serve two broad functions – orchestrating the immune response and killing infected cells. Two different classes of T cells called CD4 T cells and CD8 T cells carry out these different functions. CD stands for cluster of differentiation; it is a convention used by immunologists to refer to proteins found on the surface of cells. Biologists use the CD nomenclature to categorise different cell types – a bit like you might describe a make and model of car.

The T cells responsible for immune orchestration are the CD4 T cells, also called helper T cells. Different types of infection require different types of responses: killing a four-foot-long tapeworm requires a different strategy to killing virally infected host cells. The CD4 T cells programme the immune response to use the right package of countermeasures. They do this through a family of signalling molecules called cytokines. The vital role of CD4 T cells is clearly seen in HIV/AIDS. HIV specifically targets CD4 T cells in the equivalent of a computer virus attacking your anti-virus software. This leads to a decline in CD4 T cells and if HIV infection goes untreated the number of CD4 T cells eventually dips below a threshold where they are unable to coordinate the immune response. This opens the door to the opportunistic infections characteristic of AIDS.

The other class of T cells is the CD8 T cell, also called the killer T cell. This cell prowls around looking for infected cells. When it finds them it kills them. They are the Rick Deckard of cells,[9] testing the true from the fake and then executing the imposters.

The way that T cells recognise pathogens comes back to our friend the HLA/MHC. As you hopefully recall, the MHC displays bits of protein on the cell surface. The T cells use a complementary recognition system called the T cell receptor (TCR), which recognises a unique MHC–protein pairing. Each TCR will only recognise one pathogen and each T cell only encodes one TCR. To be protected against all of the horrible pathogens out there, you need millions and billions of different TCR molecules – estimates suggest that an average person makes 10^{13} different TCRs. The enormous diversity is generated through a genetic sleight of hand. There is only one TCR gene, but it is made up of multiple building blocks so that the cell can assemble the protein in a nearly infinite number of different ways.

IN COLD BLOOD

The final mechanism of immune protection is mediated by soluble components in the blood. It is sometimes called humoral immunity, referring back to the four humours of Galen and Hippocrates. Many substances in the blood protect us against infection, but most importantly antibody. Paul Ehrlich, a German scientist working at the turn of the nineteenth century, coined the word antibody. Together with another German scientist, Emil von Behring, Ehrlich developed a treatment for diphtheria by injecting bacterial toxin into horses. Ehrlich and von Behring later fell out due to the standard causes of scientific dispute – credit and remuneration – with Behring getting the lion's share of the credit, including the von in his name.

Antibodies are Y-shaped molecules with a recognition domain (the arms of the Y) and a functional domain (the stem of the Y). They have an incredible specificity: they only recognise one single part of one single molecule. The structure of the antibody molecule was in part determined by Rodney Porter, whom I mainly mention as one of the five winners of the Nobel Prize for Medicine who worked at Imperial College (where I work).

Like the TCR molecules, we possess a very diverse repertoire of antibodies. A recent estimate suggests that any one person can make one quintillion different antibody types (10 with 15 zeros after).[10] But rather than being made by T cells, antibodies are made by B cells.

Antibodies protect us from infection in a number of ways. They can cover the surface of the pathogen to prevent it from entering the target cell, in a process called neutralisation. Alternatively, antibodies can accelerate the removal of infectious material by making it more appetising to macrophages, in a process called opsonisation. There is also a process called complement, that we must never ever mention. The specificity of antibodies makes them an incredibly powerful tool for a range of diagnostics and therapies, of which more later. We also use different flavours of antibodies for different functions – mediated through the stem of the Y-shaped molecule. Antibodies at mucosal surfaces belong predominantly to a family called IgA. Megan Cole and Vicky Gould in my group, working with Dr Paul Turner and the team at Public Health England, explored how these molecules protect against influenza infection and whether they can be preferentially produced following vaccination.[11, 12]

T cells and B cells both have an important trick up their sleeve: memory. Following recovery from any specific pathogen, you won't get it again. This memory can be incredibly long-lived. In 1781 a measles outbreak occurred on the Faroe Islands, which was not unusual for the time, but because of the Faroes' extreme isolation they did not experience another outbreak until 1846 (sixty-five years later). Both outbreaks were severe, with a high mortality, but in 1846 individuals who had been infected during the 1781 outbreak and had survived did not get reinfected. Immune memory underlies how vaccines work.

Much of our understanding about immune memory comes from Ita Askonas. As with many of the scientists responsible for breakthroughs in understanding about the immune system and infection, Ita was born to Jewish parents in Europe but was forced to flee after the Nazi takeover of Austria in the thirties. She retained remarkable influence and was still attending Imperial College seminars well into her eighties, often asking insightful questions despite appearing to be asleep for the whole talk. I owe her a great deal, as do many others in the field, and scientists packed her funeral, despite it being in February, in the freezing snow, in North London.

Twenty-five years after reading that introductory book on immunology, I am still fascinated by it and still learning. Trying to pick the key points and squeeze them down to a short(ish) chapter that made sense was a real challenge. The key lesson is that our immune system protects and nurtures us from external threats (internal too if you include cancer). When taken together, the physical, microbial and immunological barriers mean that in spite of moving in a microbial world, infection is a rare event. But, of course, infections occasionally do happen and they do not happen in isolation: they spread through human populations. As we saw in 2020 and 2021, it is the spread of pathogens that makes them incredibly disruptive. There is a branch of medicine that models how disease spreads, which is called epidemiology, and we turn to this next, starting with Dr John Snow, without whom we would know nothing.

CHAPTER 6

Epidemiology

Timeline: Early July 2020. Airbnb in Worthing, 200 m from the sea. UK lockdown one ended. Global COVID-19 cases 11,327,790; deaths 532,340.

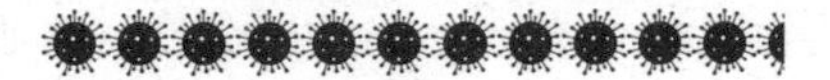

'I arrived at the conclusion in the latter part of 1848, that cholera is communicated by the evacuations from the alimentary canal.'

Dr John Snow

MODELLING HOW DISEASE spreads is vital in controlling outbreaks. In 1854 cholera emerged explosively onto the streets of London. This wasn't particularly uncommon. London is and was a hub for infection; all those people living on top of each other, not smiling, inevitably accelerates the spread of infections. However, rather than blaming bad air, a doctor named John Snow (no, not that one) traced the source of the infection back to a single pump on Broad Street (in Soho, London). In so doing, Snow began the formal study of the incidence, spread and control of disease.

One of the biggest lies of education is the need to study Latin to be able to understand other stuff. I have always thought this to be bollocks. Why study Latin when you can study the actual subject? Apparently Latin helps you understand the roots of some words; for example, the scientific names for species. I guess this is marginally helpful. The Linnaean name for chicken, *Gallus gallus domesticus*, derives from *gallus*, the Latin name for a cockerel, with *domesticus* indicating domestication. But who needs to know stuff when you can access Wikipedia, without which I wouldn't be able to tell you that the word Gallus also means Gaul, as in Asterix and chums: hence the French rugby team use a cockerel as their mascot. Another cockerel-related fact is that the Welsh word for cockerel is *ceilog* which, because it sounds like Kellogg, explains the presence of a cockerel on cornflakes packages. While we find ourselves on Latin bird name facts (and this is really stretching it but is genuinely worth sharing in a niche way), the Latin name for the hobby bird (a type of raptor) is *Falco subbuteo*: hence the hobby game of football with the same name. Come for the infectious disease knowledge and learn amazing facts about chickens – money well spent. None of which you would have learned conjugating amo, amas, amat ad nauseam.

If anything, we should study Greek. Science, as Maureen Lipman famously noted, is full of -ologies, most of which are derived from Greek words. 'Ology' comes from the Greek word *logia*, meaning to tell. All sorts of different stems can be added to this. For example, the term microbiology describes the study of microbes that do and don't infect people. Within microbiology, we find bacteriologists (studying bacteria), virologists (studying viruses), mycologists (studying fungi), parasitologists (studying parasites) and mixologists (overcharging for fruit-based spirit drinks).

The *demos* in epidemiology means the people and the *epi* part means on (which doesn't really make sense to me):

taken as a whole it loosely means the study of health related events in a population. It isn't restricted to infectious disease. Demonstrating the link between cigarettes and lung cancer is a form of epidemiology, as is demonstrating the lack of a link between 5G masts and anything except a better phone signal and faster data. In this book I am going to focus on infectious disease epidemiology.

AN OUTSIDER'S INTRODUCTION TO EPIDEMIOLOGY

So how does epidemiology help us model the spread of infectious diseases? The simplest model of the spread of an infection is SIR. Two Scottish scientists – William Kermack, a biochemist who had served as a pilot in the First World War, and Lt Col. Anderson Gray McKendrick, a military doctor who studied bacterial growth in India – contributed much of the underpinning framework for this approach in 1927. Their model splits the population into three types of people: S for susceptible – those who have not yet had the disease; I for infectious – those who have the disease and are spreading it; and R for recovered – those who have been infected and cannot be reinfected, because they are resistant or dead. People can move between the phases, so you start as an S and become an infected I and then a recovered R. This means the number of people who are S(usceptible) decreases over time and the number who are R(ecovered) increases (see figure 6A).

There are, of course, variants of this model, determined by the interaction between the host and the pathogen. For some reason, some pathogens don't induce such strong immune memories and it is possible to get reinfected with them. We, as scientists, don't really understand what enables reinfection. These kinds of interactions, where people can be reinfected, can be modelled using SIRS (the extra S for susceptible again). SARS-CoV-2, the virus that caused the COVID-19 pandemic, may behave in this way. Immune responses to other members of the same coronavirus family are known to wane and sporadic incidences of SARS reinfection were detected; though at the time of writing it was still too early to know how common reinfection with SARS-CoV-2 was and how long immunity lasted for. But it was at least six months according to one study and there was no reason to presume it wouldn't be longer.

Another way in which SIR doesn't completely reflect infection is when people act as carriers. These are people infected asymptomatically and spreading it to others without ever knowing they had it themselves. For example, typhoid, caused by *S.* Typhi, doesn't always cause disease: some people carry the bacteria asymptomatically. The most infamous of these is Mary Mallon, an ill-fated cook, now known as Typhoid Mary. She infected at least fifty-one people with typhoid without showing symptoms herself and was quarantined twice in her life. The first time she was reprieved, following support from the newspaper tycoon William Randolph Hearst (he of Rosebud fame). However, her second incarceration had no reprieve and she died in quarantine twenty-three years later. Given I went a bit mad after twenty-three days of coronavirus lockdown, it's pretty hard to imagine twenty-three years of it.

Spreading a disease without displaying symptoms causes particular problems from a disease control perspective. Some infections like Ebola virus require patients to be haemorrhagic (bleeding profusely) to spread the infection. Provided you isolate infected patients for these pathogens you can control the spread. Problems emerged during the West African Ebola virus outbreak because family members handled infected corpses without protective equipment as part of local burial practices, spreading the infections further. Other viruses sadly spread prior to symptoms. HIV, the virus that causes AIDS, is a striking example of this. Infected individuals can carry and spread the virus for a very long time before becoming aware that they are infected themselves. Undiagnosed spread was a major challenge during the COVID-19 outbreak. In most people, SARS-CoV-2 caused a mild infection that didn't look different to a normal winter cold (runny nose, cough, sneezes). This meant that foci of infection could establish rapidly and cases became evident only when the death rate soared. In late February/early March 2020, Iran reported a higher number of deaths relative to cases, indicating a long tail of infection and substantial undetected spread. Likewise, estimates suggest that there were 1,300 separate introductions of COVID-19 into the UK in the spring of 2020, rather than a single 'patient zero'.[1]

Epidemiology models can be crudely compressed into a single value to represent the speed of transmission of a virus in a population – R_0. For reasons best known to the politicians, in 2020 this became R.* A simple definition of R_0 is the number of new cases that will stem from a single infected individual (see figure 6B). So an R_0 of 2 means that every infected person will infect two additional people. A lot of doubling can occur before the spread becomes widely visible, especially if many of the cases are without symptoms.

* Presumably because they couldn't find the subscript shortcut key on their speak and spell typewriters to include the $_0$ (it's CTRL = if you were wondering).

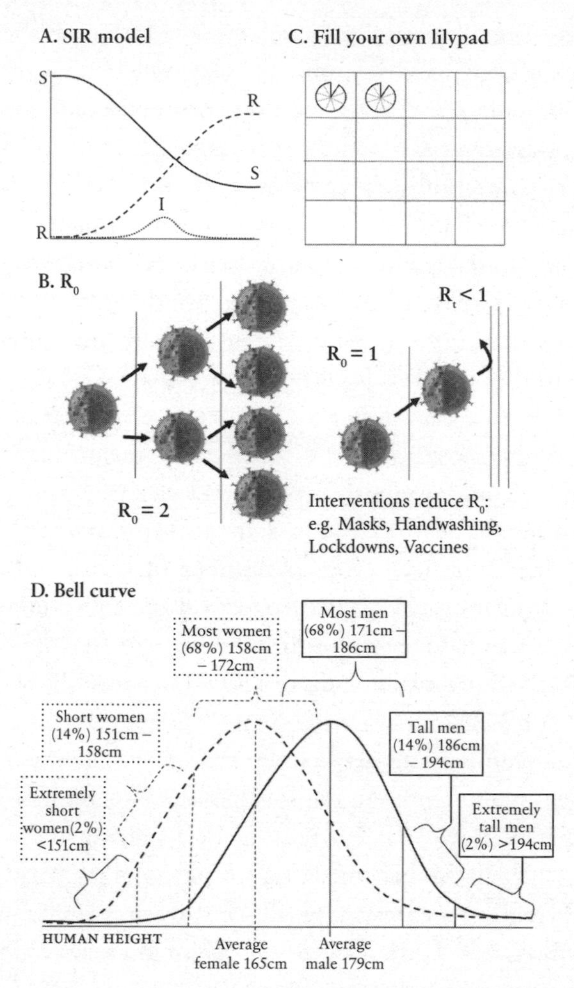

Figure 6 Some maths: A. SIR model – over time the number of susceptible (S) players decreases and recovered (R) increases. B. R_0 the basic reproduction number – showing how $R_0 = 2$ doubles but $R_0 < 1$ shrinks and this can be altered by interventions. C. Lily pad model of doubling. D. Bell curve – data demonstrating the distribution of human height; most people (68%) cluster around the average, 28% are then in the next brackets (or one standard deviation) and only 4% are in the extremes.

The time-honoured lily pad model can be used to envision this. Our imaginary, but educational, pond starts with one lily pad. The number of lily pads doubles every night, so on day two there are two new pads, on day three four new pads, on day four eight new pads and so on. Our epidemiological friends then pose the following question: if the lily pads cover the pond after thirty-six days, how many days would it take to cover half the pond?

I will carefully introduce a space here, to allow your brain to think.

Hopefully, you engaged what Daniel Kahneman calls your slow brain?[2]

If not, do so now. It is slightly a trick question.

Big clue, it isn't eighteen days. I've also included a mini lily pad so you can visualise it (see figure 6C).

With that nudge, did you come up with thirty-five days as the answer? Working backwards from 100% on day thirty-six, the day before there would be half as many lily pads, which is 50% or half the pond. Fascinatingly, until day thirty the lily pad coverage would be less than 2%.

Doubling things gets out of control quickly. There are plenty of other examples of this: folding a piece of paper, the back half of a chess board in the wheat problem or the inexplicable growth of TikTok stars. The scary thing is that for many diseases the R_0 can be even greater than 2; measles is quoted as having an R_0 of 12–18 in a naïve population (where everyone is S). If we assume the higher number (R_0=18) then in our lily pad example only 5% of the pond would be covered on the day before complete coverage. This explains why measles can spread so terrifyingly quickly – especially in clusters of unvaccinated children.

It is important to bear in mind that R_0 is an average derived from the population as a whole rather than individuals. Averages can be a misleading part of maths, especially when it comes to understanding the behaviour of biological systems. We all know that an average is the sum of the numbers added together, divided by how many numbers there are. In the sequence 1, 2, 3, 4, 5 the average or mean is 3, which is conveniently the middle value or median of this contrived series.* A lot

* Here's a poem to help: Hey diddle diddle, the median's the middle, you add and divide for the mean, the mode is the one you see the most, and the range is the difference between.

of the time we assume that values in a population distribute evenly around the average – this is called a normal distribution or bell curve (because when drawn it looks like a bell). Human height distributes normally; most people cluster around the average height of 5 ft 10 in (178 cm) for men and 5 ft 5 in (165 cm) for women. So 68% of all women are between 5 ft 2 in and 5 ft 7 in and 68% of all men are between 5 ft 7 in and 6 ft 1 in. Only 5% of women are more than 5 ft 10 in or less than 4 ft 11 in and only 5% of men are more than 6 ft 4 in or less than 5 ft 4 in. All kinds of biological outcomes fit bell curves – for example, immune response to vaccines, shoe size and weight.

However, not all sets of data are normally distributed. In the sequence 1,1,1,9 the mean is also 3, even though most of the values are 1. The number 9 in the sequence is called an outlier. Like the above sequence of numbers, the R_0 is unevenly distributed and this will be a result of variations in both human biology and behaviour. There are superspreading individuals: at a choir practice in Washington State, USA one individual infected fifty-two others in two and a half hours. There are also superspreading clusters: a biotech conference in Boston, USA, the 2020 Cheltenham Festival and Liverpool vs Atletico Madrid on 11 March 2020 in the UK are all possible events where multiple infections occurred, due to their timing early in the pandemic and the mixing of large numbers of people from several countries, though this has not been confirmed.[3] There is a very rough 80:20 rule – 80% of transmission comes from 20% of people – but this is about as accurate as any of these 'rules'.

R_0 is also not a set value. It can change over time. The value of 12–18 for measles comes from data generated between 1912 and 1979, since when patterns of human interaction (for example housing density) will have changed dramatically. Behavioural changes will also have a large impact on the actual R_0: particularly simple control measures such as handwashing, lockdowns and the usage of masks – if worn properly/at all.

R_0 will also change throughout the course of a pandemic: as an infection spreads through a population the survivors will be resistant to reinfection and so a smaller percentage of the population will be susceptible. In SIR model nomenclature, as R increases, S decreases. This adjusted R_0 is called the effective reproductive number or R_t.

Because this is a chapter on epidemiology, it doesn't seem unreasonable to include one equation:*

$$R_t = R_0 \times x$$

This means that the effective reproduction number equals the basal R_0 for a pathogen, multiplied by the susceptible proportion of the population (x). That population immunity can alter the spread of infection determines how we use vaccines (see Chapter 9).

The other important factor that affects the spread of an infection is latency. Latency is the time between exposure and the onset of symptoms – how long you go from being S to I in the SIR model. Some infections have a short latency – for example, on average it takes 1.6 days from infection to influenza. Other pathogens have a much longer incubation – it can take ten years from HIV infection to AIDS in untreated patients – and some never progress beyond latency; you can be infected with TB all your life without symptoms.

* Don't worry, there are only two other equations and they don't come until Chapter 11. Trust me, they are worth waiting for.

All of these factors can be built into a dynamic understanding of how disease will spread. The speed of spread will be a combination of latency and reproduction – in the lily pond doubling occurred every day, a combination of an R_0 of 2 and a latency of one day.

In 2020, advice from epidemiologists helped policymakers to understand the spread of an unknown disease in a rapidly changing situation. There was a lot of chatter (from 'keyboard warriors') about models and how the government should react in those chaotic days of early 2020. SAGE and the government found themselves in an impossible Goldilocks situation: there were as many people who thought lockdown was too strict as those who thought it was too lenient and all of them seemed to be on Twitter at once.

In the end, models are just models and they contain limitations. Norbert Wiener (a mathematician from the USA) and Arturo Rosenblueth (a Mexican physiologist) famously wrote: 'The best material model for a cat is another, or preferably the same cat.' Likewise, George Box (a British statistician) said: 'All models are wrong, some are useful.' Both aphorisms remind us that models do not paint the whole picture and we need to be careful not to be tied to our own personal favourite. This can be difficult when you have invested a lot of personal and reputational capital into them.

Coming back to SARS-CoV-2, the latency is four to five days and the uncontrolled R_0 is 2.6 (plus or minus). Without interventions, this can rapidly spread out of control. Here is an extremely crude, back-of-a-cigarette-packet calculation to try and demonstrate this (with lots and lots of assumptions).

> Assuming the first SARS-CoV-2 case was 27 November 2019 (an assumption made by working backwards from when the first cluster of cases emerged) and applying the R_0 and latency very crudely, the entire world would have been infected by 31 March 2020. Yet we didn't see this kind of uncontrolled exponential spread in SARS-CoV-2. For reference, the reported number of cases for 31 March 2020 was 777,798 (10,000 times fewer). The key reason for this is that the interventions introduced by health agencies, underpinned by advice from epidemiologists, *did* slow the spread of the infection, effectively reducing the R_0.

However, in order to control an infection effectively you need to know who is infected. Without some way to identify who is infected, modelling and interventions are impossible. Coming back to John Snow, he had a considerable advantage in identifying the source of the infection: because cholera lacks subtlety, tracing was easy. Other pathogens are not so easily discernible and you need tests to know who is infected. This diagnosis of infection is a core principle of controlling infectious outbreaks and the next chapter will describe some of the ways in which it is done; starting, of course, with COVID-19. So put down your home swabbing kit and read on.

CHAPTER 7

Diagnostics

Timeline: Early August 2020. Airbnb Portscatho, Cornwall, 75 m from the sea. Eat out to help out. Global COVID-19 cases 17,396,943; deaths 675,060.

'Diagnosis is not the end, but the beginning'

Martin Fischer

YOU MAY NOT remember it now, but every day of the spring of 2020 Matt Hancock (the then UK Secretary of State for Health and Social Care) would appear behind the podium, in his Pink Tie, carefully socially distanced from the actual experts (the UK very fortunately had the expertise of Professor Chris Whitty and Sir Patrick Vallance). He would then tell us about how many gagillion tests the UK had run that minute. But why should we care about testing?

It comes back to the fact that most pathogens are not visible to the naked eye. We need some way of telling who has got the infection, who has had the infection and, vitally, who hasn't. Knowing this is critical to controlling pathogen spread. Put simply, you can't do epidemiology without this knowledge, hence the need for diagnosis.

Specific diagnosis of a pathogen is not always required for treatment, especially for mild infections, or groups of infections where a similar course of action can be followed – you don't normally need to know exactly which pathogen is causing your nose to run. But there are lots of times when diagnosis is required: including when there is a new and rapidly spreading infection of concern, such as SARS-CoV-2; or a pathogen that is not self-resolving and specific drugs are required to cure it, such as HIV; or when the drugs are no longer working, such as TB.

Diagnosis identifies the causative agent of a mystery disease and in doing so stops the spread. In late 1929, a mystery cluster of pneumonia cases presented in the USA, thus beginning The Great Parrot Fever. Which I include here as much because of its wonderful name as its educational value (and the chance to make a dead parrot reference).* One of the doctors attending the sick patients had read of an Argentine actor who had become ill while playing a sailor on stage – with the obligatory Captain Flint as a prop. He referred to the disease as parrot fever in a telegram to the US Public Health Service, a name that caught on in the press, and hysteria spread faster than the disease itself. A US Navy admiral ordered his sailors to cast their parrots to the waves and another health official urged owners to wring their birds' necks. The causative bacteria, *Chlamydia psittaci*, was identified by mid-January 1930. It is endemic in members of the parrot family and can spread to humans in very close contact – for instance, if you give Polly a cracker from your mouth. But, crucially, it cannot spread from person to person. Knowing what had caused the outbreak and how it had spread led to a dramatic decline in cases.[1]

* E's not dead. E's just resting.

Historically, the only method to diagnose infection was through recording the signs and symptoms of the disease. There is a subtle difference between the two. Signs are the objective evidence of a disease as measured by someone else; for example, a temperature measured using a thermometer. Symptoms are subjective and reported by the patient; for example, feeling sweaty and feverish. I'd not appreciated this difference until I wrote my first paper and attributed some symptoms of respiratory infection to mice – while I may boast a range of talents, interpreting mouse squeaks is, alas, not one of them.

Diagnosis can vary in difficulty depending on the pathogen. Some pathogens are reasonably straightforward to diagnose: the record amount of stool produced by someone with cholera in one day is eighty litres, and while that might not give you a pathogen-specific diagnosis it is pretty clear that a person producing enough rice-water to fill a bathtub has been infected with something. Other pathogens are more subtle to detect – especially when they lead to symptoms that look like other infections. The similarity of mild SARS-CoV-2 symptoms to other winter colds contributed to the uncontrolled spread of COVID-19. If it had a unique tell – like chickenpox with its distinctive red marks everywhere – then people could have discriminated between SARS-CoV-2 and other viruses. It did present with one slightly unusual feature – some infected people transiently lost their sense of smell and taste. This led me to propose the Dutch oven diagnostic test, where you fart under the duvet in the morning to check your partner hasn't succumbed to infection. The more grown-up members of the infectious disease community widely ignored this breakthrough.

FINDING PNEUMO: HISTORY OF BACTERIAL DIAGNOSIS

Effective diagnosis of a specific pathogen only really became possible with the ability to culture (science word for grow) the pathogen itself. This links back to Koch's postulates, which showed that a specific microbe caused a specific disease. While bacteria had always been isolated from natural sources, such as van Leeuwenhoek's discoveries in pond water, Louis Pasteur was the first person to create a specific liquid for culturing bacteria, which was called growth medium or growth media depending upon your predilection for plural or singular Latin words. For his media, Pasteur boiled up meat broth, but since then various more standardised formulas have been used, which allows replication between labs – a keystone of science. Probably the most commonly used of these is called LB, which was developed by Giuseppe Bertani. Bertani originally intended LB to stand for lysogeny broth, but it is often misreferred to as Luria broth, Lennox broth or Luria–Bertani broth, which goes to show that scientists aren't always as thorough in their reading of published papers as they should be. LB contains everything the bacteria *E. coli* needs to thrive and because *E. coli* is the workhorse of much molecular biology, LB is very widely used.

If you want to make one litre of LB you will need:

- 10 g tryptone (basically milk protein)
- 5 g yeast extract (Marmite)
- 10 g NaCl (table salt)
- Dissolve in 800 ml of distilled water and cook to 121°C for twenty minutes.

This recipe was brought to you by Tom Maniatis and Joseph Sambrook, who wrote the bible of lab cloning – a hefty three-volume tome with a distinctive blue cover and white plastic ring binding.[2]

Liquid media allows us to grow large amounts of bacteria and understand how they behave but it has a relatively limited diagnostic value because you cannot count or characterise the specific bacteria in any given sample. It is therefore more useful to immobilise the media, as first demonstrated by Robert Koch. Koch initially used gelatine to set his media – the same thing that is used to make jellies for children's parties. However, in 1881 Fanny Hesse, the wife of a scientist who worked for Koch, suggested that they replace the gelatine with agar. Allegedly this suggestion occurred at a picnic where Fanny's puddings did not melt in the warm weather, though this story probably underplays her role in its development.

Agar comes from seaweed and possesses the key property of staying solid at warmer temperatures than gelatine. This is important for microbiology because most bacteria are grown at 37°C (human body temperature) and gelatine melts at 25°C: trust me, there are few things worse than a sloppy culture. Mixing your bacterial growth media with agar at a high temperature gives you a liquid that you can, like a children's jelly, pour into whatever shape you want. The most common vessel used to set the growth jelly is the Petri dish, named after another member of Koch's laboratory, Julius Petri. Petri first described the dish in 1887 – it is basically a clear, shallow dish, 10 cm in diameter and 1.5 cm deep, with a lid that covers the top, but not so tightly that air is excluded. Petri's dish was originally made from glass, but they are now plastic and microbiological labs will go through staggering amounts of them in a day. A Petri dish full of set media is referred to as a plate. It is relatively easy to make: for LB-agar add 20 g of agar (available at all fancy food stockists) to the LB you made earlier, microwave and pour. It should be noted that Petri dishes are mainly used for culturing bacteria, not viruses – so when people describe a crowded situation (e.g. the Northern Line during a busy rush hour) as a Petri dish, you can now knowingly correct them – to yourself, of course, because whatever else happens one must not talk on the Tube.

LB isn't the only media out there. A whole cookbook's worth of recipes are available – mostly named after their creators or the main components. I am a fan of Terrific Broth for its name alone. Different media support different types of bacteria; for example, some bacteria need iron or haemoglobin to grow. The easiest source of these is animal blood, the addition of which turns the liquid dark brown – hence the name chocolate media. Don't eat it, because it doesn't taste like chocolate, not even bad American chocolate.

One of the other notable media is Mueller–Hinton broth. The Hinton part takes its name from Jane Hinton, one of the first female African American vets. Before her veterinary studies, Hinton worked in a microbiology lab at Harvard with John Mueller, where they developed a media on which a range of different bacteria could grow, including the bacteria that cause gonorrhoea (*Neisseria gonorrhoeae*, which is as tricky to spell as to treat). Their media recipe contains starch, which helpfully absorbs some of the nasty chemicals made by bacteria that could otherwise interfere with any antibiotics being tested.

So how does growth media set in a small glass dish help us identify the infectious agent? Individual bacteria are too small to be seen by the naked eye. However, if you put a single bacteria onto a plate of media and leave it in the right conditions overnight, that single bacteria will grow into a colony of bacteria visible as a small round whiteish blob (colours vary with bacteria). If you forget about the plate and leave it in your incubator (an expensive oven set at body temperature) it will eventually turn into a stinky lawn of bacterial gloop.* The process of growing bacterial colonies on plates is used diagnostically; a sample is plated out (spread around a plate with a tiny hockey stick) and then left to see what grows. We can then use the plate to roughly determine the kind of bacteria. Old school microbiologists use a sniff test – different bacteria smell different – though this is now discouraged in a case of health and safety not gone mad, because sniffing potential pathogens is not the most sensible activity.

It is also possible to estimate roughly how many bacteria are present in a sample. One bacteria on a plate will form one colony, referred to as a colony forming unit (CFU). If you have 100 colonies (100 CFU in trade terms) on your plate, the original sample had 100 bacteria (see figure 7A). Problems come when the original sample is so full of bacteria that the individual colonies merge into one, making counting impossible. This can be overcome by diluting down the original sample, often multiple times, and then multiplying the colony count by the dilution. The number of bacteria in a sample can be compared to a diagnostic

* This also happens when you leave the plates in the fridge too long, often with putrescent fungal contaminants. Once in the history of science this turned out to be a good thing (see Chapter 10), but most of the time it is just rank.

threshold; for example, a urinary tract infection is defined as 1,000 CFU/ml of urine (pee normally being sterile). You can try this at home. If you had a wee now and spread 1 ml (approximately a quarter of a teaspoon) of it on the LB-agar that you made from Marmite earlier, you will hopefully find nothing. Admittedly there are a number of caveats to this – firstly, unless you work in very clean conditions, you are quite likely to get background infections from the atmosphere; secondly, I would recommend forewarning your housemates (there are few things more awkward than explaining why you are peeing on the cookware).

Nowadays, hospital labs identify bacterial infections with a newer technique called Matrix Assisted Laser Desorption Ionization-Time Of Flight Mass Spectrometry or MALDI-TOF-MS – because saying Matrix Assisted Laser Desorption Ionization-Time Of Flight Mass Spectrometry takes longer than the technique takes to run. First, the bacteria get mixed with a cocktail of chemicals that can dissolve the proteins from within the bacterial cell – the M in MALDI. Then the bacterial–chemical cocktail is bombarded with a laser which vaporises it into a gas – the ALDI. The gas then gets fired down a charged tube at a detector (called a mass spectrometer – the MS); bigger molecules will travel slower than the smaller ones – the TOF. The technique gives a fingerprint of each bacteria, which can be compared to a reference table.

FINDING DHORI (VIRUS): A HISTORY OF VIRAL DIAGNOSTICS

Spreading a sample on an agar plate is effective for bacteria but it doesn't work for all pathogens. As mentioned, viruses cannot grow by themselves: they need another organism to grow in. One approach is to use a whole animal – taking a sample from an infected person and injecting it into another species. While pioneer virologists used this approach in the early twentieth century to demonstrate the existence of viruses, it is not a sustainable, ethical or even effective approach for diagnostics.

An alternative is to use embryonated eggs with tiny chicks in them. The egg acts as a natural incubator; it contains all of the nutrients the baby bird needs to grow, which can be hijacked to grow virus. Virus can be injected into the chick and provided it is a virus that infects birds, it will grow. This method remains in use for the growth of influenza virus for vaccines. There are some limitations with this method for diagnosis, the main one being that not every human virus infects chickens. And as with liquid bacterial media, successfully growing a virus in an egg will not necessarily tell you the amount of virus in the patient sample or even what virus it is.

Therefore, alternative approaches are necessary, which is where a technique called cell culture comes in. Using a similar approach to the culturing of bacteria, animal tissues can be kept alive in a Petri dish through the provision of oxygen, warmth and nutrients. This is called *in vitro*, from the Latin for 'in glass', to differentiate from *in vivo* (in an animal) or *in silico* (on a computer) and woe betide you if you don't italicise these terms. The culture of whole animal tissues is easiest for simpler tissues – the first viruses were cultured in guinea pig corneas (corneas are basically a single cell type). Growing viruses in animal tissue was important in the initial discovery of many viruses, but it was still not practical for diagnostic labs. The real breakthrough came with the development of cell lines – the most famous of which is the HeLa cell.

HeLa is short for Henrietta Lacks. Henrietta was an African American woman born in Roanoke, Virginia, USA in 1920. In January 1951 she received the news that she had cervical cancer at Johns Hopkins hospital in Baltimore, Maryland. In order to diagnose her cancer, her doctor, Howard Jones, took a biopsy from the mass on her cervix. Unfortunately for Henrietta, cancer treatment in the 1950s was rudimentary; she received treatment with radium inserts (the standard practice at the time) but died of cancer in October 1951. Unlike most of us, Henrietta's story doesn't end with her death. Unbeknown to her, Dr Jones passed some of her biopsy to a colleague, George Otto Gey. Gey observed that the cells collected from Henrietta Lacks could be kept alive far longer than other cells. He took a single cell from Lacks's biopsy and allowed it to multiply until he had a uniform population made solely from the offspring of the original cell.

The ethics behind the HeLa cell line are complicated – while it wouldn't meet today's standards, taking and storing tissue without consent was not unusual for the time. The practice has only recently changed. In the UK, rigorous standards for the collection and storage of human tissue were first codified in law as the 2004 Human Tissue Act. The act came about in response to the Alder Hey organs scandal, which revealed the wide-scale collection and storage of human organs without consent in a number of UK hospitals. Henrietta Lacks made an extraordinary contribution to science and human health. Some attempts at reparation have been made to reflect this contribution – a large American biomedical foundation (the Howard Hughes Medical Institute) recently made a donation to her heirs.

HeLa cells, because they are derived from a single cell progenitor, can be grown reproducibly in multiple different labs – if I grow a plate of HeLa cells in my lab, they will behave the same as a plate of HeLa cells in my friend Ash's lab in Melbourne. Because they are so widely used, and essentially immortal, one source estimates that fifty billion kilograms of HeLa cells have been grown since 1950.[3] If true, that is approximately the same mass as one eighth of the world's population. In the context of viral diagnosis, HeLa and other cells are important because they can be grown on a plate – which means you can count viruses in the same way as you can count bacteria on a Petri dish. Instead of looking for globs of bacteria, viruses reveal themselves as an absence called a plaque. Cell lines like HeLa cells grow as a monolayer – if you put them into a plastic dish they will replicate until they cover the available surface to a single cell thickness. When a virus infects a cell it kills it and if the virus kills enough cells it makes a hole in the monolayer. This hole can then be visualised by the addition of a dye that binds to cells, called crystal violet. Where there are cells, the dye stains a deep purple and where the virus killed the cells colourless spots emerge. Each of these spots represents one virus – the same as each blob on an agar plate represents one bacteria. You can then estimate the amount of virus in your sample.

Viral culture is an important tool for scientists working to characterise virus – Felicity Zhang (and others) in my lab routinely grows respiratory syncytial virus in cell lines. Virus culture is also important for isolating new viruses that emerge in the community. When SARS-CoV-2 first emerged, the group at the Chinese Centre for Disease Control grew up samples in human cells to help with the identification. But the virus growth on a plate method doesn't necessarily tell you which virus is infecting your patient. The method is no longer used as a routine diagnostic because it takes time and there is some risk associated with growing up more of the deadly virus you are trying to control in the first place.

THINK LATERALLY: ANTIBODY TESTS FOR PATHOGENS

Rather than having to grow the whole pathogen, it can be identified from its constituent parts; this is faster and safer and can be done without as much specialist equipment. Current methods detect pathogen-derived proteins or genetic material.

To recognise pathogen-derived proteins we harness the power of the immune system, in particular the antibody. Antibodies are incredibly specific in what they recognise, down to individual strains of the same pathogen (see figure 7B). One common example of an antibody-based diagnostic is the pregnancy test,* which contains a lot of sophisticated biochemistry for something that you pee on (see figure 7C). The test detects a hormone called human chorionic gonadotropin (hCG), which is only made during pregnancy. When urine is applied to the test stick it moves along the material, in the same way as when you dip a biscuit in a hot drink (NB: don't mix your pee and your tea). The test stick contains a wick coated with different biochemicals in different places. The first thing the urine meets on its journey up the wick is antibody specific for hCG. These antibodies are mobile, with an enzyme attached to them that can turn a colourless chemical blue. If the urine contains hCG it binds

* Pregnancy is not exactly an infection – though it does involve the growth of a foreign organism. There is a lot of interesting immunology on why the mother doesn't reject the foetus.

to the antibody and the whole antibody-hCG complex carries on its way up the stick. The next thing the urine meets is another band of antibodies specific for hCG. Critically, more than one antibody can bind to the same protein in different places, so you can build up larger aggregations. Unlike the first antibodies, the second band of antibodies are immobilised – they are stuck in place. The immobilised antibody captures the antibody-hCG complex in a specific place on the stick (which has been pre-treated with the colour-changing chemical). We can see the presence of hCG because the enzyme attached to the antibody changes the test line to blue. The second blue line shows the test is working properly. In positive samples both lines turn blue, but negative samples only activate the second line.

The pregnancy test came on the market in 1978, building on the research of Judith Vaitukaitis, a reproductive neuroendocrinologist who isolated an antibody specific for hCG at the NIH (National Institutes of Health) in the US. Alongside the pill, the home pregnancy kit is a significant pillar of the 1970s women's liberation movement. Thumper, Mickey and Mr Toad also welcomed the test; rabbits, mice and toads were all previously used as mobile pregnancy kits (the original approach doesn't end well for them).

The same testing principle can be applied to any diagnostic protein that you can develop antibodies against; for example, prostate specific antigen in cancer. For the diagnosis of infections, the test card is coated with antibodies specific for proteins made by the pathogen of interest rather than for the hormones associated with pregnancy or cancer. The proteins are described as antigens because they are recognised by antibodies. Antigen-based rapid detection tests are used widely in the detection of malaria. During the

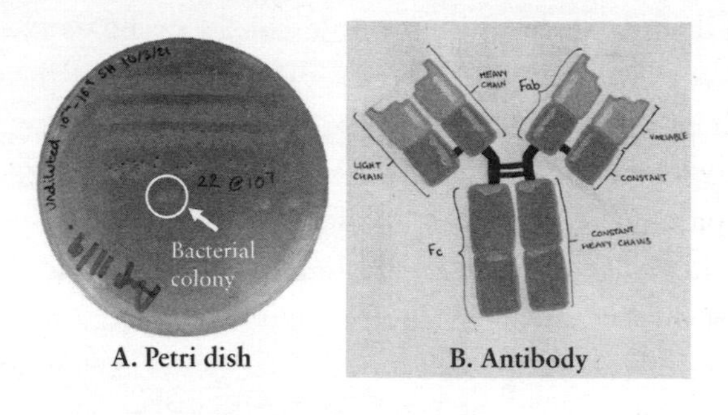

A. Petri dish **B. Antibody**

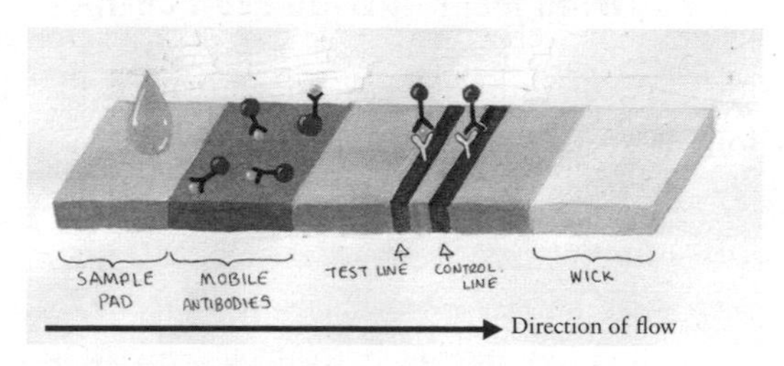

C. Pregnancy (and other antibody based) test

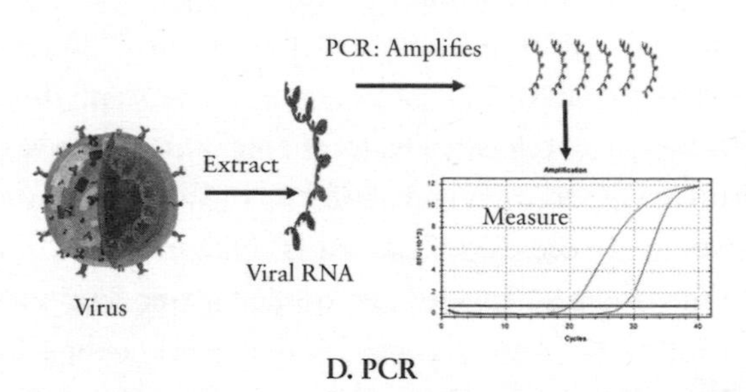

D. PCR

Figure 7 Diagnostic tests: A. Petri dish with bacterial colonies – each light circle was once a single bacteria. B. Antibody – highly specific, binds to other proteins. C. Pregnancy (and other antibody-based) tests – the sample moves up the wick, collecting antibody along the way. If the sample contains the substance of interest, it binds to the test line and causes a colour change. D. Diagnostic PCR – genetic material is extracted from a virus in a sample and PCR amplifies the material, which can then be detected.

COVID-19 pandemic these tests were called lateral flow tests (LFT). They lacked the sensitivity of PCR but were much cheaper and may have had some value in identifying asymptomatic shedders, but this was hotly debated. In the UK, parents of secondary school children will have come to hate them, because of the frequency at which we were forced to test our teenagers.

KNOWING ME, KNOWING FLU PCR-HA

The workhorse of gene-based tests is called PCR, which stands for polymerase chain reaction. Kary Mullis developed PCR while working at Cetus Corporation (which became part of Chiron, which became part of Novartis). Mullis said that he came up with the idea while driving along the Pacific Highway with his girlfriend. PCR is a staggeringly important breakthrough and Mullis got the Nobel Prize for it. However, as with a few other renowned scientists, including James Watson and Fred Hoyle, Mullis went on to develop more controversial ideas later in life. To explain why great scientists seem to go off the boil in later life, Martin Rees, the Astronomer Royal, observed that as scientists grow old they can either become an administrator, do mediocre science or strike off half-cocked into unfamiliar territory – and quickly get in over their heads.[4]

PCR is a targeted DNA amplification process. It allows us to multiply tiny amounts of DNA into large enough amounts to be visualised (see figure 7D). It works as follows. DNA, as James Watson described (before he turned out to be a massive racist[5]) is a double helix. It comprises two molecules wrapped around each other. The two strands can be separated and this normally happens during replication, when each strand acts as the template for new copies. Mullis realised that the strands can be separated by heat and that instead of making new copies of the whole piece of DNA you can make a copy of a much smaller, specific region of it.

To do this you need to target your gene of interest with two short fragments of DNA called primers, which act as bookends, telling the replication where to start and stop. You need a forward (start) primer and a reverse (stop) primer, which need to be designed based on the gene sequence. If I wanted to target the spike gene of SARS-CoV-2, I would go onto t'internet and find it at the wonderful GenBank, hosted by the NIH https://www.ncbi.nlm.nih.gov/genbank/. The spike gene has the accession code MN908947.3. Primer design can be done by hand, which is the molecular biology equivalent of rolling your own cigarettes (a fiddly affectation), but most scientists use online tools, which make the job easier.

If designed correctly, the primers will bind to the source material at the place you want it to be, and only that place. But primer design is not trivial – as can be imagined there are only so many ways in which the letters A, T, C and G can be arranged, which means you can fail to be specific enough.* If your primer is insufficiently exact it could just as easily amplify the gene for cat testosterone as it could the gene for the SARS spike protein.

Having designed our primer, we need to add a few ingredients to make more DNA: an enzyme called DNA

* It should be noted that while there are lots of visually appealing palindromes in DNA sequences, other than CAT it is quite hard to hide words in the genetic code from the sub-editors when you only have C, A, T and G to work with.

polymerase which can build new DNA molecules; the A, T, C and G themselves in a form that can be incorporated into DNA; and some magnesium. There are some profound biochemical reasons why magnesium helps, which are both beyond the scope of this book and my understanding, but if you ever find yourself in the position of setting up a new PCR reaction and it isn't working, which it won't 95% of the time because science sucks, remember this tip – add more magnesium. Alternatively try Wuffle-Dust, which Dr Ade Lennon, the hippy post-doc who trained me, used to sprinkle over my desk when things weren't working. The constituents of Wuffle-Dust remained a closely guarded secret, but it looked an awful lot like glitter. Having added our primer and the other bits to our source DNA, we can now make multiple copies of the gene we are interested in – by raising and lowering the temperature of the reaction. Each cycle of DNA melting, primer binding and DNA synthesis produces twice as many copies of the DNA of interest. To get a workable amount of DNA you need repeated cycles of hotter and then cooler temperatures. This is now done by a computer-controlled temperature block, but originally it took three water baths at different temperatures, a stopwatch and a technician named Derek.

Having come up with the process, another technical hurdle needed to be overcome for mass diagnostics – the effect of high temperatures on enzymes. Proteins don't like heat and most break at temperatures over 37°C, in a process called denaturing. This meant that in early PCR reactions, the enzymes had to be refreshed to keep the reaction going. However, in 1969 two microbiologists, Thomas Brock and Hudson Freeze, isolated a heat-loving bacteria they named *Thermophilus aquaticus* in the Mushroom Spring of

Yellowstone National Park, USA. Like all bacteria, *T. aquaticus* needs to make copies of its own DNA to replicate and to do this it uses a DNA polymerase. Unlike our puny version, DNA polymerase from *T. aquaticus* (*Taq* for short) thrives at 80°C. This was a huge breakthrough, because now the whole reaction can run without opening the tube, freeing up time for other important scientific pursuits.* In the 1980s the only place you could buy *Taq* was Roche and they made nearly $2 billion from it. Bugs are big business.

There is one final step in the process. PCR amplifies the amount of DNA in the original sample, but it needs to be visualised in some way. Fortunately, some dyes specifically bind to DNA molecules (they are called intercalating dyes, because they fit inside the helix). In diagnostic PCR, these are incorporated at each round of amplification, giving you more and more signal – and a distinctive S curve in response, as seen at the bottom of Figure 7.

PCR is now routinely used to detect viral infections. However, those of you paying close attention will remember that SARS-CoV-2 is an RNA virus, with genes made of RNA, not DNA. RNA is a single-stranded molecule – it lacks a dance partner. This is useful for its main role, which is transmitting information from the DNA in the nucleus to the rest of the cell. But it does mean that normal PCR doesn't work. Luckily, we have stolen another trick from pathogens, this time viruses. The 'central dogma' used to teach molecular biology is that information flows from DNA to RNA to protein: the DNA is the central information store; the RNA

* Or spend more time messing about on the internet. As is often the case, the time freed up by the modernisation of science doesn't always end up being spent productively.

is the information conduit; and the protein is the final prod uct. Viruses, because they spend more time in the bars and nightclubs of universities than the lecture theatres, don't always obey this rule. One family of viruses, called the retro-viruses, encode their genes on RNA but use an enzyme called reverse transcriptase to turn this back into DNA (more of which in Chapter 11). Using reverse transcriptase, we can turn RNA into DNA prior to plugging it into our PCR machine to see if it is from the virus of interest.

As with all things, PCR can be misused or misinterpreted. Because PCR amplifies the input material, any small mistake will also be amplified. The mistakes mostly come from contamination in the sample preparation steps – the 'crap in, crap out' problem. A normal PCR reaction has forty cycles, which therefore could produce 2^{40} (two times two, forty times over) copies of the DNA molecule, assuming it doubles each cycle. There have been numerous incidences where people 'discovered' viruses and attributed them to conditions lacking a cause – falling down on the 'causation ≠ correlation' problem.

The story of XMRV (xenotropic murine leukaemia virus-related virus) and chronic fatigue syndrome illustrates the problems associated with PCR misdiagnosis. Chronic fatigue syndrome is a complex condition, the main symptom being extreme tiredness, and the cause is not fully understood. In 2009, a group led by Dr Judy Mikovits published a (since discredited) paper that linked chronic fatigue syndrome to XMRV. Multiple independent studies had been unable to repeat Mikovits's finding and the journal retracted her paper for its flaws.[6] Dr Mikovits has a chequered history – the institute she worked at when she published her findings filed a suit against her for removing laboratory notebooks and she was briefly imprisoned.[7] As discussed in more detail in Chapter 14, bad research such as this poisons the well, misleading individuals with the disease. Following the study, the poor people suffering from chronic fatigue syndrome were given hope that there might be a viral cause of their condition and therefore potentially a cure, only to discover it was not true. Unfortunately, rather than target the researchers who got it wrong, some affected individuals chose to target the researchers who disproved the association.

Another area for concern with the use of PCR was the suggestion that SARS-CoV-2 RNA could be detected in sewage water: one study hypothesised that virus was circulating as early as March 2019.[8] But this study was not conclusive; they only found a positive response at a high cycle number, which should always be interpreted with caution because of the doubling nature of PCR. The other concern is that SARS-CoV-2 is not the only coronavirus that can infect humans (the 2 in the name being a pretty big giveaway); as such it will share gene sequences with other co-circulating coronaviruses. The final problem is the lack of concurrent disease – given the result was from a large volume of sewage water (800 ml) there should have been a large peak of viral transmission in the community at the time, which was not seen. This makes me sceptical of this finding. It is important to note that at the time of writing (this book) the findings about SARS-CoV-2 in wastewater had not been through the full scrutiny normally associated with scientific publication – they were made available as an unevaluated pre-print. I will return to the problems of science by press release in Chapter 14, but it's safe to say that some results aren't as robust as others.

All of which is to say that PCR has its limitations; but it is nonetheless a powerful diagnostic tool and very widely used. PCR allows us to tell whether our sample contains virus. Even more powerfully, it can be multiplexed – multiple viruses can be identified in the same sample. Which means for approximately the price of a Starbucks coffee we can determine viral infection in a sample in less than twenty-four hours.

READING OUR GENES

Another gene detection tool exists that gives an even *greater* level of detail about the pathogen: this is gene sequencing, where you read off the letters that make up the DNA (or RNA) of an organism. Fred Sanger developed the original methodology in Cambridge, UK. Sanger was no slouch. Before he started his work on DNA, he developed a methodology for reading the sequence of proteins – identifying the exact amino acid blocks that make them up. For this he won the first of his two Nobel Prizes; putting him in very exclusive company. The other double science winners are John Bardeen, who invented the transistor and developed superconductor theory; Marie Curie, who on top of being a science bad-ass discovered both radioactivity and radium; and Linus Pauling, who explained how atoms bond together and campaigned against the nuclear bomb – for which he got the Peace Prize. Two of the PhD students who worked in Sanger's lab went on to get Nobel Prizes for their work (Rodney Porter for antibodies as mentioned earlier and Elizabeth Blackburn for her work on telomeres), which is probably more remarkable – there are quite a lot of geniuses in science, but far fewer are good mentors as well.

Having cracked proteins in 1952, Sanger moved on to DNA. He published the method for sequencing DNA in 1975. His method, now known as Sanger sequencing, uses an approach called chain termination. Two principles underpin the method: sequential synthesis and size separation. DNA synthesis occurs sequentially, one letter added at a time: if you imagine the letters like pieces of a children's train track, then normally each piece connects to the letter before and after. Returning to the description of PCR earlier in the chapter, you might have noticed the sequence was written 5′ CAATTG 3′. The 5′ and 3′ orientate the biologist to the order of the sequence: they refer to specific positions on the nucleotide building blocks and how they join together. The 5′ position of one nucleotide (a phosphate group) clicks into the 3′ position (an oxygen atom) of the preceding nucleotide.

To continue the train analogy, as well as continuing the track you can add dead ends to finish it off. The same can be done using chemical analogues of the DNA nucleotides (the letters that make the DNA), which will stop the synthesis – terminating the chain. If you mix a cocktail of normal DNA nucleotides and the chain terminating ones, you will get DNA of different lengths.

We can now separate these using gel electrophoresis – applying an electrical current through a great big jelly. This separates the DNA fragments by size: different-sized molecules move through a gel at different speeds, with smaller ones moving faster and therefore further; this size separation can achieve resolution down to individual nucleotides.

Finally, we need to visualise the different length DNA fragments. Historically, this was done with radioisotopes: radioactive versions of atoms that make up the DNA, which can be seen with photographic film. Sanger sequencing runs four reactions: each one contains three of the four DNA nucleotides as normal molecules which can extend the chain and critically one specific letter is a radioactively labelled chain terminator. So in the A (adenosine) reaction tube you will end up with a mixture of different lengths of DNA, all of which end in the letter A; in the T (thymidine) tube all will end in T, etc. The four reactions are run side by side through a gel and will give you a ladder pattern that you can read across from one base to the next. Three warnings if you set out to try this at home: 1) nucleotides run towards the red electrode – if you plug the gel in the wrong way around the radioactive gloop goes out of, not into, the gel; 2) the gel is made of a highly toxic substance called acrylamide, which sets only if you add a second chemical, TEMED – if you forget to do this, you end up with a toxic radioactive soup on your lab bench; 3) the tips used to load the mix into the gel are really bendy – if you catch them on the glass plate while loading the gel you fire a fine spray of radioactive oops across the lab (allegedly). The good news is that coloured dye has now replaced radiation and most labs now outsource to an external contractor anyway.

Sanger's method was the backbone of sequencing for the next thirty-five years. Once developed, researchers raced to sequence increasingly complex organisms. The first organism sequenced was the bacteria *H. influenzae*, followed by the yeast used in baking and brewing, *Saccharomyces cerevisiae*. The first multicellular organism sequenced was the nematode worm, *Caenorhabditis elegans*, followed in quick succession by a plant, *Arabidopsis thaliana*, then the fruit fly and then the mouse. This culminated in the Human Genome Project, which in 2004 published a draft of all of the genes that make a person

– at the cost of $2.7 billion or approximately one US dollar per letter (it is 3.2 billion base pairs long).* Clearly, at that price it would not be practical to use sequencing for diagnosis. However, in parallel with Moore's law (which states computing power will double every two years), the cost of sequencing has drastically dropped over time. It is now possible to sequence an entire human genome for $1,000. The 100,000 Genomes Project in the UK took less than six years to reach that benchmark and is an invaluable resource for understanding the genetic underpinning of disease.

Gene sequencing is not routinely used diagnostically, but it is very effective for tracking the spread of infections within a community. While PCR tells you whether a person is infected, sequencing allows you to track exactly which variant of the virus they are infected with. This enables maps of geographical spread to be constructed – a form of molecular epidemiology. A friend of mine, Professor Ian Goodfellow, used this approach during the West African Ebola outbreak between 2013 and 2016. In December 2014, he upped sticks from his lab in Cambridge, UK and moved to Makeni, Sierra Leone to establish one of the first diagnostic labs in the country. Using modern techniques, he could sequence a whole virus in twenty-four hours and then use that data to inform contact tracing. This was not without risks – Ian normally works on norovirus, which is unpleasant but not deadly, but Ebola virus is an entirely different kettle of fish. Wildlife added a level of complexity not normally found in labs in the leafy Cambridge suburbs – Ian and his team found a deadly spider in the containment

* It was only fully completed in 2021 – there are long, repetitive stretches at the end which haven't been read; a bit like footnotes.

lab and ants in the robot used to dispense liquids: the ants managed to get into the experimental mix. All told, viral sequencing directly contributed to the eradication of Ebola in Sierra Leone. Building on this experience, he and others developed a deployable sequencing lab that fits into a suitcase. To make it this small, Ian used portable hydroponics greenhouses; the issue being that hydroponics greenhouses are often used for less noble purposes. Ian told me that explaining to Dr J-J Muyembe, one of the discoverers of Ebola, why the sequencing labs had pictures of cannabis leaves stamped on them was somewhat awkward!

During the COVID-19 pandemic this form of molecular epidemiology developed further. Multiple groups worked together on this, as befits a global pandemic, mapping the spread of the virus in real time and sharing results in both conventional scientific publications and on social media. This data has suggested the single emergence of the virus (from an animal reservoir, not 5G masts or meteors). The family trees of the virus also make it possible to track roughly how many times a virus entered a country or region. Another UK-based consortium (COG-UK) had sequenced over 100,000 viral genomes by 10 November 2020.[9] This genome level diagnosis was critical in tracking the evolution of the virus. It was how the new British (Alpha), South African (Beta), Brazilian (Gamma) and Indian (Delta) variants of concern, which spread faster, were identified.

LET YOUR BODY DO THE TALKING

All of the above methodologies directly look for the pathogen itself, but the immune response can be used as a flag for infection, past or present.

The main immune component studied is antibody. Because of our immune memory, our blood tells a personalised Bayeux Tapestry of our own battles with infection. In theory you could map a person's entire history, though that would probably need all of their blood and cells, which rather defeats the point. For instance, following infection with SARS-CoV-2 you should have antibodies against SARS-CoV-2 in your blood. Antibody and infection history do not always perfectly marry up – it is possible to have had an infection without symptoms and develop antibodies detectable by the tests, and it is possible to have had an infection with symptoms and not develop antibodies; it is, however, impossible to develop antibodies to a pathogen without ever being infected.* The lack of alignment between infection and measurable antibodies may also come down to the sensitivity of the test used.

Measuring antibody in blood is called serology, which is named after blood with the cells spun out of it, called serum. Serology can be used to track the spread of an infection; for example, isolated populations can be used to understand transmission. Since few things are more closeted than an English public school, following the 2009 swine flu pandemic Public Health England measured antibody responses at Eton College and compared them to records of who had been ill. While three quarters of the students had positive antibody response to the virus, only a quarter had been to the sick bay, suggesting a big iceberg of asymptomatic patients. As seen in COVID-19, asymptomatic patients make tracking an infection very difficult, because they can spread the virus without ever knowing they had it.

* Except after vaccination (which is kind of the point).

Antibodies are so specific that they can identify differences down to individual strains of a pathogen. The different strains of a pathogen that can be differentiated by antibody are called serotypes. The bacterial pathogen *Streptococcus pneumoniae*, the predominant cause of community-acquired pneumonia, has over ninety different serotypes. This means that an antibody that recognises *S. pneumoniae* strain 1 will not recognise *S. pneumoniae* strain 2. There are so many *S. pneumoniae* strains because the composition of the bacterial coat changes (specifically the sugars). We will look at the impact of this on vaccine development in Chapter 9.

Rebecca Lancefield performed much of the early work on bacterial serotyping while working at Rockefeller University in the early twentieth century. Lancefield worked on β-haemolytic streptococci. She developed a taxonomy of grouping these bacteria based on the sugars found on their cell walls. Microbiologists still use Lancefield's taxonomy; it runs from A to S. This includes group A streptococci or *Streptococcus pyogenes*, the cause of scarlet fever, and group B streptococci or *Streptococcus agalactiae*, the cause of severe disease and meningitis in newborn children.

Antibodies are not the only diagnostic immune marker; T cells can also be used as a diagnostic tool. Using a technique called an ELISpot (enzyme-linked immune absorbent spot, if you really wanted to know), the number of pathogen-specific T cells in a blood sample can be quantified. This approach is used to diagnose TB infection, specifically latent TB.

Inhaling *Mycobacterium tuberculosis* leads to a range of outcomes. The first is nothing at all – you exhale it back out or it gets caught in mucus and is removed. The second outcome is full-blown TB, which is relatively easy to spot – as John Keats recognised after finding blood on his handkerchief.* However, there is a third outcome – latent infection, where the bacteria enters cells and replicates but doesn't cause disease. Nearly a quarter of the world's population has a latent TB infection according to some estimates[10] and since 10% of people with latent TB go on to develop the full disease this is clearly concerning. Detection of latent TB infection early can prevent the need to resort to romantic poetry. The standard way of detecting latent TB used to be a combination of a chest X-ray and the tuberculin skin test. The tuberculin skin test is pretty crude: the patient is injected in the forearm with a bolus of proteins from *Mycobacterium tuberculosis*. It was developed by none other than Robert Koch and refined by Florence Seibert, an American biochemist who identified the active ingredient in the mixture, making it much more reliable. Patients with TB develop a swelling at the injection site within two to three days, but the skin test suffers both from false positives (anyone who has had the BCG vaccine) and false negatives (anyone infected less than ten weeks earlier). For latent TB the tuberculin skin test has been replaced by the ELISpot assay because of its greater sensitivity (if it is positive you definitely have TB) and greater specificity (if it is negative you definitely don't).

The TB skin test isn't the only test with issues associated with incorrect results. Two main types of statistical error

* 'I cannot be deceived in that colour;– that drop of blood is my death-warrant;– I must die.' Hospitals rely on a combination of chest X-ray, culturing of the sputum for bacteria and a form of PCR called GeneXpert, rather than poetic tendency.

exist: false positives and false negatives. I will use the example of COVID-19 PCR tests to illustrate this. Having made yourself gag by swabbing your tonsils and then posting the tube to a testing station, you waited for a result. This would either come back positive, when viral RNA had been detected, or negative, when no viral RNA was detected.* With a perfect test, every positive result would mean that the person definitely had the virus and every negative result would mean that the person did not. However, no test is perfect: the COVID-19 PCR test was about 95% accurate when used on people with mild symptoms. The sensitivity increased when used in conjunction with the known symptoms of COVID-19 – fever, new continuous cough and loss of smell. False positives were an issue when they meant that people had to isolate unnecessarily, but on the whole it was better to err on the side of caution in order to reduce the spread of a highly contagious virus.

Returning to COVID-19, Matt Hancock and his pink tie, in April 2020 he promised to deliver 100,000 tests a day. To be fair to him, Public Health England (PHE), the body responsible for public health in England (obvs), *did* deliver 100,000 tests by the end of April 2020, which increased to 200,000 tests a day in May 2020 and peaked at 801,949 on 3 February 2021. The majority of these tests used PCR. This reflects a massive undertaking and a lot of the media coverage at the time failed to refer to what a remarkable effort it was. For reference, I feel pretty happy with myself when I have done one successful PCR in a day. To deliver 200,000

* There was a third result – inadequate. I got this for one of my tests and at the time it felt as much a comment on my life in 2020 as it did on my COVID-19 status.

tests a day requires a lot of the right machines – in March and April the British Army helped requisition PCR machines from research labs across the land; which was a strange feeling, knowing your lab equipment was contributing in a time of national crisis (a bit like giving your old pans to build Spitfires). Mass testing also needs a lot of competent people to be working very hard, very carefully on what is quite a repetitive task. Fortunately, many people heard the call – including students and staff from my department, such as Akshay Sabnis, Anne-Marie Levins and Dr Mike Skinner. Credit should also go to Rachael Quinlan, Yasmin Mallu and Aileen Rowan, who provided me and my team with testing kits throughout the pandemic as part of an asymptomatic screening programme.

The invisible enemy is no longer invisible. This enables the epidemiologists to do their modelling and the public health experts to apply control measures. Of course, you will only find the virus if you actually look for it. In July 2020, the then president, Donald Trump, suggested: 'When you test, you create cases.' The same muddle-headed logic was trotted out by COVID-19 denialists when the UK genomics consortium first identified the more rapidly transmissible variant in autumn 2020. Detection rates do not always tally with prevalence and not looking won't stop the spread.

I have now given you an overview of the pathogens that threaten our health and the science underpinning how we fight them. In the second part of the book I will describe how science and technology prevents, treats, cures and eradicates infectious diseases.

PART 2

Solutions: Preventing and curing infectious disease

CHAPTER 8

Prevention

Timeline: Mid-August 2020. Isle of Wight, 300 m from the sea. France removed from safe corridor, prompting mass exodus. Global COVID-19 cases 21,026,758; deaths 755,786.

'An ounce of prevention is worth a pound of cure'
Benjamin Franklin

IN THE SECOND half of the book I am going to talk about human interventions to control infectious disease, starting with the Black Death; which we now know to be caused by the bacteria *Yersinia pestis* and not a great miasma associated with a conjunction of the planets. Fleas spread it and as every schoolchild knows the fleas were carried on our old friend *Rattus rattus*, the black rat. However, recent research partially exonerates the rat. The Black Death first emerged in gerbils in the deserts of Kazakhstan. It was then transmitted to traders working the Silk Road, via human parasites such as lice.

Wherever it came from, the Black Death entered Europe in 1347 at the port of Caffa (now Feodosija) in Crimea when, in one of the first recorded incidences of biological warfare, Tatar troops launched plague-infected bodies into the city.[1] Infected traders fled the city and spread the disease to the Italian ports of Genoa, Venice and Pisa and from there it spread by land, affecting almost all of Europe by 1353 – except Poland, curiously.

In the response to the Black Death, we can see the beginnings of infection control, built on the key foundation of stopping people ever becoming infected in the first place. This was not completely effective in the 1300s – the connection between bacteria, flea and disease had not been made.* However, the Middle Ages authorities at least recognised that infection could spread from person to person; it felt as if some people in 2021 had lost this knowledge. Improved understanding of the causative agents of infectious disease and how they transmit led to better strategies to control spread. Nevertheless, some of the approaches used during the Black Death to control infection were still in use, unchanged, seven hundred years later, during the COVID-19 pandemic.

CONTACT LOST: QUARANTINES, LAZARETTOS AND LOCKDOWNS

The simplest way to prevent the spread of infection is to stop contact between people. This comes in two forms – keep all the infected people out or keep all the infected people in. The principle of separating infected people (or quarantining) is recorded as early as the book of Leviticus

* By which I mean basically useless, with nearly half the population of Europe dying.

in the Bible. The word quarantine itself links back to the Black Death in fourteenth-century Venice and means forty days. Initially the period was thirty days, so we could have known the isolation period as a 'Trentine' instead, but the Venetian senate felt thirty days was not long enough. The quarantine principle has been applied ever since – the astronauts on Apollo 11 were quarantined after their return, just in case they had contracted something on the moon.

Quarantine and isolation played a major part in the SARS-CoV-2 containment strategy. Size and connectivity determine the effectiveness of this approach. New Zealand, an island with a relatively small population (five million, about half the size of London) effectively pulled down the curtain in March 2020, applying a fourteen-day quarantine to all travellers entering the country, followed by a total lockdown on 23 March. The country was declared COVID-free on 1 May 2020. New Zealand remained basically COVID-free from that point on. One of the more jaw-dropping statistics was that following an event at the Rose Garden in September 2020 (with no social distancing or mask wearing), there were more cases of SARS in the White House than in the *whole* of New Zealand. Different countries tried different approaches – in the UK it was a bit of a lottery as to which countries you were and weren't allowed into at any particular time. At the time of writing in 2020, I was OK to go to Finland and Norway, but not to Sweden.

The alternative to closing borders is to put all your infected people in one place. During the Black Death, maritime cities such as Venice, Dubrovnik and Marseille had isolation areas called lazarettos. Lazaretto is a portmanteau word combining Lazarus, the leprous beggar in the Bible who made a miraculous recovery from death, and the island of Santa Maria di Nazaretum. Leprosy has historically been associated with the isolation of infected individuals in leper colonies; for example, Devil's Island off the coast of French Guiana. It is caused by the bacteria *Mycobacterium leprae*, which is from the same genus as the bacteria that causes TB. But rather than infecting lung cells, *M. leprae* infects nerve cells, leading to the characteristic loss of pain sensation and the subsequent damage to extremities associated with leprosy. As a highly visible and contagious condition with no drugs to treat it, isolation was one of the few strategies available.

One breakthrough came in the 1910s, when Alice Ball, the first female African American professor at the University of Hawaii, developed a treatment. Ball, a chemist, started with Chaulmoogra oil from the *Hydnocarpus wightianus* tree; the oil could treat *M. leprae* but it was very difficult to use, because it was gloopy and tasted disgusting. She developed a method to extract the active compounds from the oil. The method was highly effective and led to the complete recovery of seventy-eight patients at Hawaii's leper colony. Ball sadly died in 1916 and in a depressingly common story of misappropriation another chemist at the university took her results and published them without giving her credit. Thanks to historians of science, the record has been redressed and Ball's role in the discovery of the treatment has become more widely known. Chaulmoogra oil was the first drug for leprosy. It was eventually replaced by treatment using a cocktail of antibiotic drugs (dapsone, rifampicin and clofazimine). While the drugs are not very pleasant, they are effective. In 2018 there were 209,000 cases of leprosy worldwide, down from 5.2 million in the 1980s, making leprosy another good news story of modern medicine.

Containment is still in use but it is no longer at the scale of the leper colonies or for the length of time that 'Typhoid' Mary Mallon endured. The Royal Free Hospital in London has a high-level isolation unit, where one of the British nurses who contracted Ebola, Will Pooley, was treated. In a testament to Pooley's courage, he returned to Sierra Leone to treat patients two months after recovering himself. Containment played a key role during the first SARS outbreak in 2003, caused by SARS-CoV-1.* SARS-CoV-1 mainly spreads from patients after symptoms emerge, making isolation of individuals more effective. Carlo Urbani, an Italian doctor working in Hanoi, first recognised SARS 1. Urbani perceived that the infections he saw in February 2003 had a novel cause and through his swift action and notification to the WHO he triggered the containment of the disease. Tragically, Urbani contracted SARS himself and died in March 2003. As was seen again in 2020, healthcare workers put themselves in harm's way to treat infectious outbreaks, often paying the ultimate price themselves.

Public health agencies and governments introduced a variant of separating people to prevent the spread of infectious disease during the COVID-19 pandemic – the lockdown. Just in case you don't remember this (maybe you blanked it from your mind), people were restricted to their homes with the occasional prison yard breaks for exercise. The virus started spreading globally in January and February 2020 and by early April 2020 3.9 billion people (more than half the global population) found themselves

* SARS-CoV-1 is not to be confused with its more successful younger sibling, SARS-CoV-2.

under lockdown in one form or another. Lockdowns have been previously used in the context of public health and safety – following the Chernobyl nuclear catastrophe ninety thousand people were evacuated from a 30 km radius around the plant, and following the 9/11 terrorist attack on New York City the whole of Manhattan was isolated. The COVID-19 lockdowns in 2020 were the first wide-scale usage of the tool in the context of infectious disease. They were, where applied and followed, successful – the number of cases (and the R_0 number) did, after about three to six weeks, fall in most areas under lockdown.

So why use lockdown for SARS-CoV-2 and not other pathogens? Three factors contributed: severity, speed of spread and scientific progress. SARS-CoV-2 is a respiratory virus spreading by airborne droplets, causing lung infections – in this respect it is not dissimilar to influenza. However, compared to influenza SARS-CoV-2 has a higher infection fatality rate (IFR): for every one thousand people infected between five and ten people will die. This is ten times higher than either the 1956, 1968 or 2009 influenza pandemics, necessitating more drastic control measures.

But there are other infections with an even higher IFR that haven't led to mass lockdowns. SARS-CoV-1 has a ten times higher IFR than SARS-CoV-2 (10% compared to 1%) and Ebola kills 90% of those infected. So, the second reason lockdowns were used for SARS-CoV-2 was the speed and invisibility of transmission. SARS-CoV-2 can spread before people become symptomatic, making it extremely difficult to trace. The lack of pre-existing immunity accelerated its spread. Most people have seen some variety of influenza, so some basal immunity exists which may slow transmission down.

The final reason is improved scientific understanding. The closest comparator is the 1918 influenza pandemic, which killed fifty million people; by way of comparison, the First World War, which ended in 1918, led to twenty million deaths. The 1918 pandemic is sometimes called Spanish flu, not because it first emerged in Spain but because Spanish newspapers first reported it. Newspapers from the combatant countries censored coverage of the emergence of the virus, including the USA where it was first detected, but Spain as a neutral country reported cases, hence the name. The 1918 virus had a 2.5% fatality rate – even higher than SARS-CoV-2 – and there were pockets where the impact was even greater, including the Pacific Islands; for example, in Western Samoa 30% of adult males died. Eastern Samoa, by contrast, imposed an early quarantine and there were no deaths on the island. The difference in response between 1918 and 2020 reflects increased understanding about how infections spread and better global communication in infection control. By the end of January 2021, slightly more than a year after it had emerged, SARS-CoV-2 had claimed 2,200,000 lives; a tragedy, but nothing in the realms of the 1918 pandemic – this is in part because of better infection control measures.

GOOD BEHAVIOUR

Quarantines, isolations and lockdowns all come with considerable social, economic and mental health costs. Where possible, other approaches are used to prevent the spread of infection between people. One measure that became common parlance in 2020 was social distancing or, more accurately, physical distancing. The principle behind this is that droplets of virus fall out of the air over a certain distance, so if we all stay in two-metre diameter bubbles we will not pass or receive infections. The principle is sound enough but it was not always fully understood. Passing someone in the street less than one metre away was probably less risky than stepping off the pavement into the path of an oncoming bus. It is not clear what impact distancing had; I think it was probably more sociology than virology – a means of increasing awareness and altering behaviour. A fine line exists in public health messaging – infectious diseases need to be scary enough that people do sensible things like wash their hands, get vaccinated, use antibiotics appropriately and invest in my research, but we don't want them to be so scary that people buy all the loo roll.

Two other behavioural interventions became much more common in the UK (and elsewhere) during the COVID-19 pandemic: handwashing and face masks. Of the two, handwashing was considerably less controversial, though this was not always the case. Prior to the 1860s, surgeons were not the meticulously clean individuals we think of today. They took pride in the number of bloodstains on their surgical gowns, as a mark of the operations that had gone well. Unsurprisingly, the rate of post-operative infections was very high indeed. Ignaz Semmelweis, a Hungarian doctor working in Vienna, first challenged this dogma: he linked the practice of examining corpses prior to delivering babies with a high rate of fever in mothers after birth. Semmelweis introduced a policy of handwashing, which led to a dramatic reduction in maternal deaths. Sadly, the establishment at the time rejected his ideas and the practice did not catch on until the work of a second surgeon, Joseph Lister (from whom we get Listerine). Lister had the benefit of working after Pasteur's publication of the germ theory, so he had a more tangible threat to control. Drawing on the work of Pasteur and Semmelweis, Lister began to explore chemical approaches to reduce infection. He developed the first antiseptic, successfully preventing infection through the use of a carbolic acid-soaked wound dressing in 1865. This work led Lister to a wider programme of cleanliness in the operating theatre, including handwashing, gloves and sterilising instruments, thereby changing the course of operating forever.

A contemporary of Lister's was Florence Nightingale; most famously remembered as the 'The Lady with the Lamp' for her service in the military hospital at Scutari (Istanbul) during the Crimean War in the 1850s. She initiated a programme of sanitation reform at the hospital, leading to a reduction in mortality. In the context of infectious disease, her greater contribution came as a pioneer of the statistical analysis into causes of death. She developed a form of pie chart (the Nightingale Rose Diagram) that recorded how soldiers in the British Army in India died over time which, as with her role in Scutari, led to better hygiene and far fewer deaths from infection.

Hygiene and handwashing are still a remarkably effective means of preventing the spread of disease. Handwashing has been shown to reduce the transmission of diarrhoea by 40% and colds by 20%. It is reasonably easy to see how handwashing works for gastro-intestinal infections – breaking the hand to poo to hand to mouth chain of transmission. For respiratory infections the role of hands in transmission is a bit more subtle – the chain goes something like: cough into person one's hand, hand to surface (for example, pole in Tube train), person two's hand to surface, hand to face. We touch our faces about twenty times an hour[2] and if we have been unfortunate enough to pick up a stray droplet of something nasty on the way that can lead to infection.

Infections can be picked up from any surface, making handwashing an incredibly important health intervention. The role of infectious droplets and the route of transmission of SARS-CoV-2 were unclear. While the virus could be detected on different surfaces under lab conditions, infectious virus proved much harder to detect from surfaces in the real world.* Dr Jie Zhou, working with Professor Wendy Barclay at Imperial College, spent a lot of 2020 vacuuming the air and swabbing the handrails in Tube trains, but never found the virus. As Dr Seuss might ask:

Can I catch it in a plane?
Can I catch it on a train?
Can I catch it at my home?
Could I catch it from a gnome?
Can I catch it at work?
Would I catch it from a fork?
What if I wear a mask?
Who knows, you'll have to ask.

For such a simple intervention, handwashing is not widely used. Globally, the rate of handwashing with soap after going to the toilet is only 20%, most of which comes down to access. However, prior to COVID-19 in the UK, where access to soap and water isn't an issue, the frequency of

* I'd say live, because that's how I think of infectious virus, but as mentioned before, they are not technically alive.

handwashing was embarrassingly low – one study has reported that only 32% of men washed their hands after going to a toilet in a motorway service area, with faecal bacteria found on 25% of people's hands. Thankfully, the percentage of people washing their hands increased during 2020.[3] One thing I noticed in March 2020, at the last concert before every venue shut down, was a lengthy queue in the gent's toilets. Strikingly, the queue wasn't for the stalls; it was for the sinks. This increase in handwashing will hopefully be a positive legacy of the pandemic. Unfortunately, I foresee a time in the not-too-distant future when we return to helping ourselves to piss-covered Cornish pasties at service stations.

Speaking of contaminated food – I am sad to report that the five-second rule has no actual basis in science. Bacteria can transfer from a surface to food in less than a second.[4] Before you get too cross that science has ruined that Percy Pig you just dropped on the floor, transfer to gummy candy was the lowest, especially if it landed on a carpet.

Hand in hand with scrubbing during COVID-19 came the obsession with sanitiser; of the two, proper washing with soap and water is more effective,[5] though sanitiser was a good stopgap in the absence of access to sinks, e.g. while travelling. The volumes of sanitiser produced were staggering, with universities and distillers producing their own brands to make up the shortfall. One of the legacies of 2020 will be finding half-used bottles of hand sanitiser squirrelled away in homes, cars and offices for years to come.

A considerably less popular intervention was face masks. It didn't help that there was scant evidence that masks are effective in the real world – most of the published studies had looked at certified face masks used in hospital settings.

There is a huge difference between effective PPE (personal protective equipment) that has been validated by the manufacturer and worn by someone who has been trained to do so and a knitted scarf. Data from the Norwegian Institute of Public Health suggested that 200,000 people would need to wear a mask to prevent one case in the real world.[6] Another trial, with a lot of limitations, indicated that mask wearing has limited impact,[7] a problem being that many people do not use masks properly. You only had to go to any enclosed public space in the UK in the second half of 2020 to see masks over the mouth only, or just the chin or sometimes only one ear, if at all; in German they were called *Maskentrottel* (mask-idiot, someone who wears their mask under their noses).

Cultural attitudes determine mask usage, which is very common in Japan, being derived from a culture of not wanting to infect other people. In the UK and the USA there was much more resistance: and not for the first time. In 1918 the Anti-Mask League of San Francisco formed, in response to an ordinance requiring people living in the city to wear masks. Similar protests, encouraged by the internet, took place in 2020, predominantly fuelled by civil liberty concerns, internet bilge and mixed messages from some politicians.

While each individual intervention may only have had a small impact, collectively they did have a big effect, particularly on other non-COVID-19 infections, which fell dramatically. In the first seven weeks of 2021, PHE did not detect a single case of influenza in the UK. It is of great interest that approaches such as handwashing, distancing and masks (sometimes called Non-Pharmaceutical Interventions or NPI) had a greater impact on other infections than COVID-19.

This may reflect the greater infectivity of SARS-CoV-2 or some underlying basal immunity. Disease modellers suggest that we might see delayed peaks of influenza or RSV as underlying immunity wanes. Ultimately, it is hard to separate what works from what doesn't: one model I liked described interventions as slices of Swiss cheese – each one has holes but if you put enough together you will get complete coverage.

Handwashing and masks won't work against all infections, because not all infections spread by coughs, sneezes or faeces. Behavioural interventions can also influence pathogens spread by other routes, blocking the pathogen spreading from one person to another; for example, condoms to prevent the spread of STIs or needle exchange programmes to reduce the spread of blood-borne diseases. Chemical barriers can also be effective. PrEP (pre-exposure prophylaxis) can prevent HIV transmission in high-risk individuals, the anti-HIV drugs acting as chemical condoms, killing the virus before it infects. Education and access underpin many of these interventions – an intervention only works if people actually use it, properly.

DRAIN THE SWAMP

Education and access are reflective of the most potent tool in infectious disease prevention – increasing prosperity. As European countries industrialised in the nineteenth century and modernised in the twentieth century important changes occurred that dramatically increased life expectancy, including different land usage and reduced use of wood fires for heating and cooking. Malaria, for example, was common in the UK between the 1500s and 1700s and was known at the time as the ague. The disease declined in the UK following the drainage of mosquito-infested salt marshes. The marshes were drained to provide more agricultural land rather than reduce malaria, because the role of mosquitoes as vectors had not yet been determined.

Mosquitoes and malaria were not linked until the end of the nineteenth century, when Sir Ronald Ross joined the dots. Ross was a child of the Empire. Born in India and then educated in Britain he joined the Indian medical service in 1881. In 1897 he fed mosquitoes on the blood of an infected volunteer before dissecting them (the mosquitoes, not the volunteer). In the stomachs of the dissected mosquitoes he found malaria parasites, thereby demonstrating their role as vectors. Brilliantly, he celebrated the result by writing a poem, including the lines:

Seeking His secret deeds
With tears and toiling breath,
I find thy cunning seeds,
O million-murdering Death.

And I think it is a shame that scientists no longer write poems to celebrate their science. The closest I have got is doggerel in the form of science Valentines:

Roses are red
Kittens are fluffy
Antiviral Immune Response is
a Trigger of FUS Proteinopathy.

Which is probably why they don't!

Draining swamps prevents yellow fever, another disease spread by mosquitoes, though this is not the same species of mosquito – there are over 3,500 different mosquito species. Malaria is spread by mosquitoes of the genera *Anopheles* and yellow fever by the genera *Aedes*. Yellow fever causes damage to the liver, leading to the characteristic jaundice, hence the name. It was particularly prevalent during the digging of the Panama Canal, until Walter Reed, an American army major, identified the link between the mosquito and yellow fever transmission. Reed initiated a programme of swamp drainage and clearance, which eliminated mosquitoes in the canal zone and dramatically reduced yellow fever during its construction.

Reed's successful programme in Panama set in motion other programmes to reduce mosquito-borne infections. Mosquitoes were successfully eradicated in the United States in the 1940s and '50s through a combination of wetland draining and use of the insecticide DDT (dichloro-diphenyl-trichloroethane). Mosquito eradication programmes were then attempted on a global scale. While these were successful in southern Europe, unfortunately DDT resistant mosquitoes emerged, rendering the insecticide ineffective. The programme stalled in the late 1970s, with a resurgence in malaria cases.

However, the good news is that eradicating malaria has returned to the global health agenda and between 2000 and 2016 the number of countries with endemic malaria dropped from 106 to 86, with malaria deaths falling by 60%. Unfortunately, less progress has been made in the last five years, the programme having been set back by both the Ebola and COVID-19 outbreaks. One concern is that with climate change, habitats supporting infection-carrying mosquitoes may alter, increasing the number of countries where the disease is endemic. That said, on 25 February

2021 El Salvador was the first Central American country to be certified as malaria-free by the WHO,[8] demonstrating that local leadership and international support can lead to significant changes.

CLEAN THE WATER

Preventing malaria by draining swamps is one approach by which changing the environment can reduce exposure to pathogens. But the environmental change with the biggest impact is clean water, as seen in the UK in the 1850s. The Metropolis Water Act 1852 had improved the water supply into London but had done nothing about effluent. Even John Snow's successful identification of contaminated water as a source of infection in 1854 didn't lead to change immediately. In the end it required someone to kick up a stink in Parliament; more specifically, the Great Stink. In the summer of 1858, heat aerosolised the contents of the Thames, at that time effectively an open sewer.* The river ponged so badly that Parliament approved the plans for Joseph Bazalgette to build 1,100 miles of sewer. This project, once completed, effectively ended cholera in London.

Clean water has had a huge global effect on life expectancy: between 1900 and 1928, the number of typhoid cases in the USA fell from 100 cases per 100,000 people to 33.8 cases per 100,000 people, thanks to improved sewerage, and by 2006 this rate was 0.1 per 100,000 people (with most cases in travellers).[9] Sadly, access to clean water is still not universal: 2.2 billion people lack access to safe drinking water.[10] Clean water for all has been targeted by the UN as

* Some people might argue that things haven't changed that much.

one of their seventeen sustainable development goals to be achieved by 2030.

Public health interventions significantly reduce infectious disease. Simply put, prevention is better than cure, both in terms of health and economics: one study estimated that every £1 spent on public health saves £14 in return.[11] Tellingly, the same study, published in 2017 before the COVID-19 pandemic, concluded: 'Cuts to public health budgets in high-income countries therefore represent a false economy and are likely to generate billions of pounds of additional costs to health services and the wider economy.'

But sometimes prevention is problematic, because preventing infection in the first place means people don't see it. The Black Death left no doubt that an infection was spreading through the population – piles of bodies, mass graves and red crosses on doors will focus the mind that way. If you can't see something happening, it becomes easier to believe that the interventions are not necessary. We see this in the resistance to fluoridation to prevent bacterial tooth decay, in spite of its significant long-term health benefits. But nowhere is this more the case than with vaccines, which alongside clean water are our most potent tool for preventing infection. In the next chapter, I will cover a brief history of vaccines and how they work. I will also explore vaccine hesitancy and why in part this is a price we pay because vaccination works so well.

CHAPTER 9

Vaccines

Timeline: Early September 2020. Back in Epsom. UK schools reopen. Global COVID-19 cases 26,121,999; deaths 864,618.

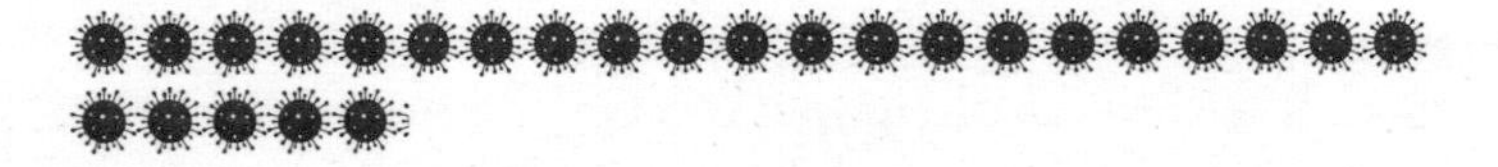

'Life or death for a young child too often depends on whether he or she is born in a country where vaccines are available or not.'

Nelson Mandela

SMALLPOX INFECTED 10–15 million people in 1967, the equivalent of every single person in London. Since 1979, no one has died of smallpox anywhere in the world, or even been infected with it. Putting that into context, in the same forty-year period more people have been killed by lava lamp explosions, suffocation after an atomic-wedgie or run-ins with sacrificial goats. The reason more people have died with their pants stuffed up their butt cracks than smallpox is down to vaccines, one of the most important human health interventions of the last two hundred years.

At this point, as an immunologist I am obliged to tell you about Edward Jenner, who was born in 1749 in a rural community in Gloucestershire, UK. One of his formative experiences was undergoing variolation, an approach developed in fifteenth-century China to prevent smallpox. It involved blowing dried-up scabs into the nostrils of a previously uninfected person; but not fresh scabs, because that would be gross.* Variolation mostly caused a mild smallpox infection, which would prevent future disease. Lady Mary Wortley Montagu, a smallpox survivor, introduced the process to the UK. As the wife of the British ambassador to the Ottoman Empire, she had observed the practice in Istanbul and in 1718 she had her son inoculated. On returning to the UK, she worked with Dr Maitland, the embassy doctor, to demonstrate the efficacy of the process to the Royal Family. They did this by inoculating seven death-row 'volunteers' at Newgate Prison, who were offered their freedom in return for variolation – luckily for them it worked. The Royal Family adopted the practice, greatly increasing its popularity.

However, variolation was not without risk; for example, Prince Octavius, the thirteenth child of King George III, died following inoculation. This wasn't overly surprising with 'the scab' not being a precise unit of measurement. This risk of disease led Jenner to look for a safer alternative. The story told by his biographer is that Jenner heard a milkmaid boast of her immunity to smallpox and therefore lack of scars, which led him to investigate the link between cowpox and immunity. The truth is more prosaic. Another local doctor, John Fewster, had made observations linking cowpox infections with protection against smallpox and

* Dried scabs are also slightly less infectious.

Jenner had probably heard these ideas as part of Fewster's social circle.[1] While less romantic than the tale of the 'beautiful milkmaid', Jenner developing his idea from other people's previous work rings more true; science grows stepwise, building on what came before. Jenner's contribution was to test Fewster's idea with an experiment. In 1796 he inoculated James Phipps, the son of his gardener, with cowpox from a milkmaid and then tried to infect him with smallpox twenty times. It's not recorded what Phipps Senior thought of this or what level of consent Phipps Junior gave. Luckily for young James the vaccine worked and as well as receiving the first reported vaccination in history, he also earned himself the free lease on a cottage. Jenner repeated the study on twenty-three other subjects, including his own son, and following this success he developed a national and international programme of vaccination. As well as developing the science of vaccinology, he played a key role in its etymology: he coined the word vaccine from *vacca*, the Latin word for cow, and in recognition of this the cowpox virus was named vaccinia. As an aside, the virus that causes smallpox is called variola, after the Latin for the pustules it causes.

Remarkably, Jenner developed vaccination sixty years before Pasteur developed the germ theory of disease and one hundred years before the isolation of the first human virus. Jenner's vaccine came from a combination of observation and testing: he and others had observed the protective effect of cowpox and then tested the theory – this evidence-based, or empirical, approach underpins much of the history of vaccine development. Basically, testing things that might work and using the ones that do. As our understanding about both the pathogens and the immune system increases, there are more observations to be made and more hypotheses to be tested, leading to more vaccines.

A BRIEF HISTORY OF VACCINES

The history of vaccines is coupled with the history of biomedical innovation. The first vaccines developed targeted bacteria, because they were easier to isolate and grow. Vaccines against viruses had to wait until we could isolate viruses, which in turn required the development of cell lines. And the innovation continued. In order to fight bacteria that hide themselves in a sugary coat, vaccinologists had to work out how to get the immune system to recognise sugars, which they did by adding some protein – a bit like hidden veg in children's meals. But some bacteria hide in a coat of the body's own sugars. To get around this problem vaccinologists used a brute force approach; harnessing the gene sequencing revolution, they tested every single gene until they found a combination that worked. However, these approaches all take time and some pathogens emerge much faster and necessitate vaccines to be developed more rapidly: the next step used platform technologies to prevent rapidly emerging pandemics.

Let's now look at the history of vaccines in more detail, because of their tremendous impact and the way in which they illustrate the interplay of innovation and inspiration. I will group them chronologically using Figure 8 as a template.

ATTENUATION (BACTERIA)

After Jenner had made the initial breakthrough showing that you could induce protection against an infection, the

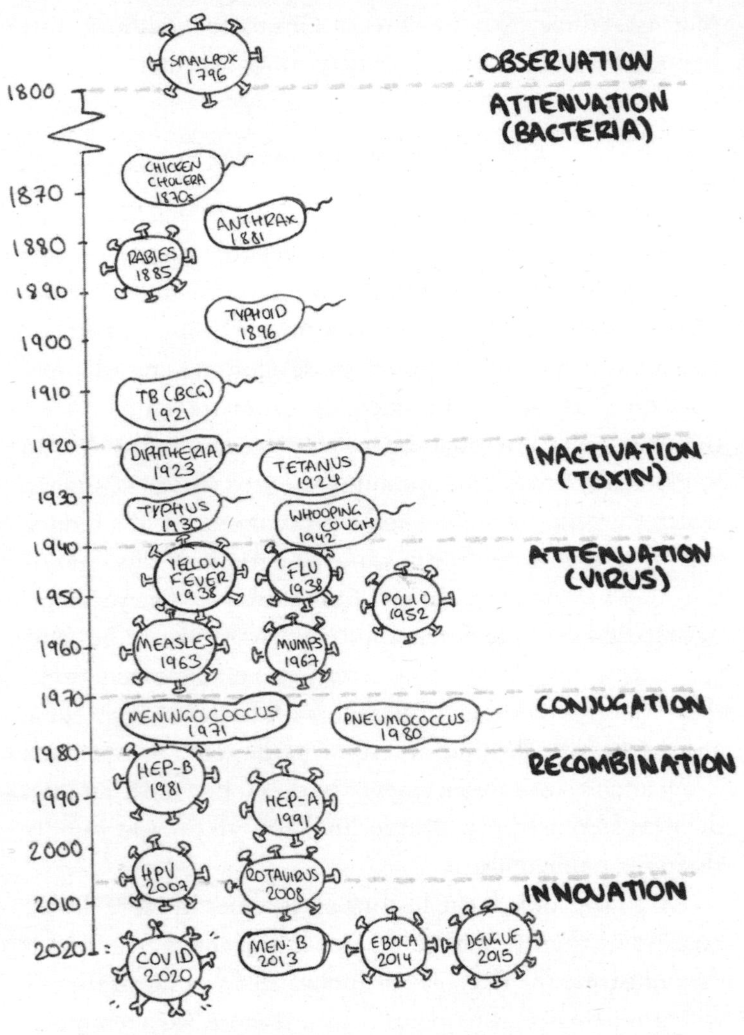

Figure 8 A vaccine timeline: Vaccine development has come in waves, reflecting breakthroughs in technology – the bacterial vaccines preceded the viral vaccines because of the order in which they were discovered. Viral vaccines required the development of cell culture (e.g. HeLa cells). The development of recombinant proteins in the 1980s led to further breakthroughs.

next big jump in understanding came after Pasteur, Koch and Cohn had linked infection to bacteria. Pasteur built on this knowledge to develop vaccines. The first vaccine he developed prevented cholera in chickens. In the 1870s Pasteur received a culture of bacteria from Jean Joseph Henri Toussaint (so good they named him thrice), but the culture went off and Pasteur failed to infect any chickens with it. When Pasteur went to reinfect the chickens with fresh bacteria they still failed to get sick. As luck would have it, Pasteur could repeat these studies using another batch of weakened bacteria, because his lab assistant, Charles Chamberland, had forgotten to infect some chickens before he went on holiday and had allowed his culture to go off.* When Chamberland returned from holiday, no doubt fearing Pasteur's wrath, he did the experiment with the spoiled culture and saw no infection. At this point Chamberland wanted to bin everything and start from scratch, but Pasteur intervened and demonstrated protection following inoculation with the weakened strain. The causative bacteria was subsequently named *Pasteurella*, though it might have been fairer to name it after Toussaint, as he had first isolated it.

The process of deliberately weakening a pathogen to make a vaccine is called attenuation. The word comes from the Latin to thin and was originally applied to eating dry figs to reduce body volume, explosively. Attenuation is subtly different from Jenner's approach, which used an entirely different virus to induce protection.

* Chamberland was not entirely hapless, because he developed the Chamberland filter, a porcelain filter with holes so small no bacteria can pass through. It was with this type of filter that Ivanovsky first isolated viruses.

Pasteur also used attenuation to make a rabies vaccine, though not without controversy. Rabies is a viral infection passed by animal bite, with a very high mortality rate if untreated. The rabies virus infects the nervous system, moving up to the brain; a side effect of the infection is hydrophobia, making it difficult to drink water. Pasteur made his rabies vaccine by infecting rabbits with the saliva of rabid dogs and then isolating their nervous systems. As with Jenner, Pasteur tested his vaccine on an unwitting boy, Joseph Meister, who had been mauled by a rabid dog. Meister did not develop rabies and went on to work in the Pasteur Institute. As well as being good news for Meister this outcome was fortunate for Pasteur, who intervened at some personal risk – he lacked a medical licence, so would have been prosecuted had Meister not survived.

Pasteur then turned his attention to anthrax, caused by *Bacillus anthracis*. It is primarily a disease of sheep and cattle, spread by long-living spores which persist in the soil until a suitable host comes along. Anthrax can infect humans and has a bit of a dark history as a bioweapon. The Japanese army in Manchuria first used it in the 1930s on Chinese prisoners of war. Soon after this, everyone got in on the act. The British government initiated an anthrax bioweapon programme in the Second World War to disrupt German livestock. They tested the dispersion of the agent on Gruinard Island in Scotland, inadvertently rendering it uninhabitable for nearly fifty years. The lengthy lifespan of the spores influenced the decision not to use it during the war. Even so, both the Americans and the Soviet Union produced anthrax as a bioweapon until at least 1972 and the signing of the Biological Weapons Convention. However, an accidental leak in 1979 in the town of Sverdlovsk showed that the USSR was still producing weaponised anthrax. The deadly legacy survived into the twenty-first century. In 2001, a week after the 9/11 plane strikes, several anthrax-containing letters were sent to US media outlets and two senators, leading to five deaths. The FBI linked these to a single source, Bruce Ivins, who worked at the US biodefence labs at Fort Detrick, Maryland.[2]

Amongst all the great things that Pasteur did, the anthrax vaccine is the most problematic. Pasteur claimed he used oxygen to attenuate the bacteria in the vaccine. It is, however, pretty likely that Pasteur first demonstrated a chemically killed vaccine, developed not by him but by Toussaint – the same scientist who had sent him the chicken cholera culture. Pasteur's guilty secret remained hidden until after his death, because he stipulated in his will that his notebooks should remain secret for 100 years. Why he needed to steal the approach from Toussaint (if he did) remains unclear, though presumably he had a monumental ego and couldn't stand to lose face. There were also financial considerations – if he had admitted to using Toussaint's approach, he would not have profited from the development.[3]

Around the same time, while working at St Mary's Hospital in Paddington, London, where I work now, Almroth Wright developed a vaccine against typhoid, used by the British Army in the First World War. He killed the bacteria with chemicals in a way similar to Toussaint/Pasteur's anthrax vaccine. Wright also set up the labs where Alexander Fleming later discovered penicillin. But as with many of the scientists after whom research labs are named, Wright's legacy outside the lab is more challenging. He was deeply opposed to women's suffrage, writing a polemic against it in 1913[4] which questioned women's intellect, public morality and physical force. While it's natural to be judgmental when hearing this today, as a product of the nineteenth century these views were not unique to Wright; it is also nice to speculate whether, if he were resurrected, he would be more upset that his old department is now run by a brilliant woman or that Fleming is considerably more famous. Probably a toss-up.

The demonstration that you can deliberately weaken a bacteria to make a vaccine led to the development of many other vaccines – for example the TB vaccine still in use today, BCG. BCG stands for Bacillus Calmette-Guérin; it is named after Albert Calmette and Camille Guérin, who developed the vaccine by performing 239 rounds of subculture over thirteen years, each time selecting for a weaker strain of the bacteria. They finally produced an avirulent strain in 1919.

INACTIVATION (TOXIN)

Pasteur, Wright and others based their vaccines on whole microorganisms. The next breakthrough built on the observation that protection could be mediated by part of the pathogen rather than all of it. Ever since Behring and Ehrlich had first discovered antibodies (see Chapter 5), blood from horses injected with bacterial toxins had been used to protect against bacterial infections. This treatment is called passive immunisation because you protect people with someone else's immunity. While effective in the short term, it does not give you a lasting protection against future infections. A similar approach remains in use today; for example, convalescent sera (blood from recovered individuals) was used as an experimental treatment for both COVID-19 and Ebola.

Horse antisera is no longer used, because it contains proteins foreign to the human immune system and can lead to a type of allergic reaction called hypersensitivity.

However, at the time of the First World War there was little alternative, resulting in a big market for antisera. For example, the drugs company Wellcome produced a diphtheria antitoxin. While working on this antitoxin, Alexander Glenny developed the first vaccine targeting a single bacterial component. Diphtheria, as well as being tricky to spell, is a respiratory infection caused by the even harder to spell bacteria *Corynebacterium diphtheriae*. The bacteria infects the airways and leads to a thick accumulation of dead cells, making it hard to breathe. The cells die because diphtheria makes a deadly protein (diphtheria toxin) which inhibits the ability of the human cells to make more proteins. It is potent stuff, with a lethal dose of 0.1 μg per kg body weight.* In an average adult about 7.0 μg would be lethal; which is less mass than two eyelashes or one grain of sugar. Glenny had been making the antisera by injecting horses with diphtheria toxin, but at some point around 1925 he discovered that he could inactivate the toxin by treating it with a chemical called formalin.[5] Proteins need to be flexible to be functional. Treating them with chemicals breaks this flexibility and therefore the functionality. Denny realised that immunising with inactivated toxin could lead to an immune response without the deadly side effects. He called his inactivated toxin a toxoid and the approach remains in use for both diphtheria and tetanus vaccines.

Chemical treatment can also be used on whole bacteria. Pearl Kendrick, Grace Eldering and Loney Clinton Gordon

* One μg is one microgram or one millionth of a gram.

worked together in Michigan to develop a vaccine against whooping cough. Whooping cough, caused by the bacteria *Bordetella pertussis*, is named because of the sharp intake of breath children take after a coughing fit. Working with an extremely small budget in the middle of the Great Depression, Kendrick and Eldering performed wonders, experimenting late into the night because there was no funding for basic research. They developed better methods for growing *Bordetella* and using a strain that Gordon had identified they developed an inactivated vaccine. Their vaccine was rolled out across America from 1939, leading to a 150-fold decline in cases.

INACTIVATION/ ATTENUATION (VIRUSES)

All of which meant that by the end of the 1930s there were several bacterial vaccines but only two viral vaccines (rabies and smallpox). This reflects the time gap between the discovery of bacteria and viruses. Max Theiler developed the first culture-derived viral vaccine that targeted yellow fever. Theiler used the multiple passage attenuation approach, similar to that used for BCG, growing round after round of the virus, first in mouse tissues and then in chickens. Somewhere between the 89th and 114th repetition, Theiler isolated a virus that had lost some of its infectious properties. Remarkably, he then continued with another 110 repetitions to ensure its safety. The virus vaccine that Theiler isolated – called YF 17D – is still used today, with over four hundred million doses having been produced. It is remarkably effective, giving lifelong immunity from a single dose.

Theiler developed his yellow fever vaccine using animal tissues, but the golden age of viral vaccines came with the development of cell lines. A major limiting factor had been the inability to grow viruses in bulk: cell lines changed this significantly. HeLa and other immortal cells were vital for the development of not one but two polio vaccines. Poliovirus infection causes poliomyelitis, which leads to paralysis in 0.5% of cases. Polio infected many famous people in the twentieth century, including Donald Sutherland, Joni Mitchell, Neil Young and even Tarzan himself, Johnny Weissmuller. Aside from Ian Dury (of the Blockheads), who ended up with a paralysed arm, one of the most influential people to catch polio was Franklin D. Roosevelt, who caught it as an adult and was subsequently paralysed from the waist down. FDR's infection led to the March of Dimes, a public subscription campaign that funded much of the vaccine research.

To develop the polio vaccine, two scientists raced for the good of all mankind – Jonas Salk and Albert Sabin. The rivalry, though famous in vaccine circles, amounted more to handbags than fisticuffs. Salk and Sabin had two opposing views of how to make a vaccine: Salk preferred an inactivated virus and Sabin an attenuated one. Salk, though younger, won the race with an inactivated vaccine grown in HeLa cells. Following a successful trial in nearly two million children in 1954, Salk was declared a miracle worker and received widespread applause for his work.* Sadly, this acclaim was somewhat premature – in 1955 a batch of the vaccine made at Cutter Laboratories was incorrectly inactivated, leading to forty thousand cases of polio and five deaths.

* Though not from Sabin, who described him as a 'mere kitchen chemist'.

Sabin continued to develop his live attenuated vaccine in parallel and finally generated one by 1954. Unfortunately, Sabin couldn't test it in America because most children had already received the Salk vaccine, which probably doubled down his frustration. Instead, Sabin tested his vaccine in the USSR. A similar challenge with vaccine testing occurred during the race to develop a COVID-19 vaccine. Once one vaccine candidate is successful, it becomes a lot harder to trial other candidates – who wants to be in a trial with a 50% chance of getting a placebo when you can definitely get the real working vaccine?

While historically Salk and Sabin received much of the acclaim, another vaccinologist deserves much more praise and recognition than he receives – Maurice Hilleman. Hilleman is the most important person from the twentieth century you have never heard of. Early in his career, while working at the Walter Reed Army Medical Center (named after the yellow fever pioneer), Hilleman predicted the severity of the 1957 flu outbreak and initiated the manufacture of a vaccine, reducing the impact of the pandemic in the USA. Hilleman spent most of his career working for Merck. Between 1957 and 1984 Merck developed forty new vaccines, many of which remain in use today. One of them is the vaccine for mumps. Mumps virus infection causes swelling of the salivary glands and it can lead to viral meningitis and male infertility. It is extremely uncommon now, thanks to Hilleman's vaccine – named Jeryl Lynn after his daughter, because he isolated the virus from her throat. This virus strain is still included in the MMR vaccine. It is very hard to quantify the true legacy of Hilleman, but he saved hundreds of millions of lives and prevented even more people from suffering from disease and disability.

CONJUGATION

Up to the 1970s, most vaccines targeted proteins in one form or another, either inactivated like diphtheria, in a whole dead virus like influenza or in a whole live virus like mumps. However, not all pathogens are covered in proteins. Some bacterial pathogens have a sugar coat, including both *Streptococcus pneumoniae*, a cause of pneumonia, and *Neisseria meningitidis*, a cause of meningitis.* Sugars are poorly immunogenic: the immune system is bad at remembering them. If you recall from Chapter 5, immune memory resides in two types of cells: T and B cells. To get the best immune memory you need both T and B cells to recognise the same pathogen, a bit like the two-man rule for nuclear weapons launch. Unfortunately for vaccinologists, sugars are invisible to T cells. This means that even though B cells can see sugars, they don't get the additional help from T cells, thereby weakening the memory.

Fortunately, there is a way around this conundrum. Work done in the 1920s had shown that sticking a sugar to a protein improves the response. This is because the T cell recognises the protein part and the B cell the sugar part and since the protein and the sugar are conjoined, the T and B cells can now talk to each other and launch an immune response. Glycoconjugate vaccines apply this approach to target bacteria: sticking sugar from the outer coat of a pathogenic bacteria to a protein and thus improving the immunogenicity. Rachel Schneerson and John Robbins developed the first vaccine using this approach, against *Haemophilus influenzae* type B, in 1987. Other vaccines

* Microbiologists once again proving their flair for naming things.

soon followed, dramatically reducing the rate of childhood meningitis.

The development of any vaccine is only half of the story; it still needs testing and rolling out. Much of the credit for the successful deployment of the sugar-based vaccines belongs to the public health authorities in the UK. They have had various guises: the Health Protection Agency (HPA) was replaced in 2003 by Public Health England (PHE). Who knows, by the time you read this it may have another name. Whatever it comes to be called it has done an amazing service, working closely with clinicians and academics to perform large trials to demonstrate the potency of vaccines and continues to do so in the age of COVID-19.

RECOMBINATION

Hilleman also laid the foundation for a vaccine against hepatitis B (HepB), a viral infection of the liver that can cause cancer and cirrhosis (long-term damage of the liver – often seen in alcoholics). Hilleman took blood from infected individuals and isolated a hepatitis-derived viral protein called hepatitis B surface antigen. His inactivation process was safe and effective but not practical for large-scale development – there is only so much human blood that can be farmed for viral protein. However, his work laid the groundwork for the next generation vaccine by demonstrating that hepatitis B surface antigen could protect against infection if only an alternative manufacturing approach could be developed.

Fortunately, vaccine innovation has always been in step with other advances in biomedical sciences. The next revolution, recombinant technology, changed the way therapeutic proteins were made. Many proteins find use as medicines – for example, insulin to control type 1 diabetes. Before the 1980s, insulin came from pig pancreases, but new technologies enabled the manufacture of human insulin in bacterial cells. Proteins, as you now know, are made by cells from RNA messages which are encoded by genes in DNA. Using tools such as restriction enzymes (the ones bacteria use to chop up foreign DNA to stop it infecting them), it became possible to cut the gene from one organism – in this case humans – and stitch it into the genome of another organism. Genentech, an American biotech company, used *E. coli* in 1982 to produce the first recombinant insulin approved for human use. Pablo DT Valenzuela, working at another California-based biotech company, Chiron, saw the potential for this approach for vaccination. Valenzuela cloned the hepatitis B surface antigen, the basis of Hilleman's blood-derived vaccine, into yeast. This yeast-derived vaccine was licensed in 1986.

INNOVATION

There followed a slight hiatus in vaccine development. When I started to lecture students in 2005, it had been ten years since the release of the most recent vaccine, hepatitis A, another Hilleman special. Having seen so much progress in the twentieth century, the lack of progress in the twenty-first had led to some concern that vaccines had hit a wall. There were worries about costs and times of development. However, in the last fifteen years there has been a renaissance of vaccines. This directly overlaps with the time I have been lecturing about vaccines – I'll leave you to interpret that how you wish.

The next technological breakthrough built upon the genetic sequencing revolution: the same revolution that changed the face of diagnostics. While polysaccharide vaccines had an enormous effect on the number of cases of bacterial meningitis, one of the bacteria that caused meningitis, *N. meningitidis* serovar B, resisted vaccination by this approach. The sugar on its surface resembles a human sugar so closely that any vaccine targeting it might train the body to attack itself. An alternative method was needed, which is where sequencing came in. A team led by Rino Rappuoli, at what was then Chiron, subsequently Novartis and now GSK (drugs companies, like health protection agencies, change names more often than Liz Taylor changed husbands) made a breakthrough in the beautiful Tuscan hills surrounding the medieval city of Siena. In close association with Craig Venter, a pioneer of the Human Genome Project, Chiron sequenced the whole genome of serogroup B *N. meningitidis* and made every single protein with the same recombinant technology used for the HepB vaccine. All 350 proteins were tested in combination to see which ones gave the best protection and eventually just four were settled on and combined into a vaccine, which was licensed as Bexsero in 2015.

ACCELERATION

Normally vaccines take time to develop. Bexsero took twenty years and the malaria vaccine Mosquirix took thirty. What if time is in short supply?

This question came into sharp focus during the West African Ebola epidemic between 2013 and 2016, the same outbreak that had Ian Goodfellow chasing spiders out of his sequencing tent. There had been other Ebola outbreaks before 2013. The first confirmed outbreak was in Zaire (now the Democratic Republic of the Congo) in 1976, near the Ebola river, from which it took its name. The initial identification of the virus has been attributed to a sample of blood collected from a Belgian nun by J-J Muyembe and sent to Peter Piot, who characterised the new virus. The story, as always, is a bit more complicated, with multiple players working across several countries, but Muyembe and Piot were definitely involved in the discovery.[6] Either way, it started a long and distinguished career for Piot, who was also critically involved in the early days of the African AIDS epidemic. After 1976 there were several relatively small Ebola outbreaks in the DRC, but the 2013 outbreak was different in both scope and scale. It spread across three countries in West Africa: Sierra Leone, Liberia and Guinea, with sporadic cases in Mali and Nigeria, and at least 28,616 people were infected, leading to 11,310 deaths.

The size of the outbreak led to an international effort to develop a vaccine and at speed. During the outbreak, my wife, Dr Charlie Weller,* worked at the Wellcome Trust, the offshoot of the Wellcome drugs company where Alexander Glenny made the first diphtheria vaccine. Wellcome coordinated the efforts to test a vaccine and my wife's role was somewhat accidental; she happened to be in the office one summer morning when Professor Sir Jeremy Farrar, the head, passed through and asked for a little help with the Ebola project. Around 100,000 air miles later she was part of an international task force working with the WHO to

* Confusingly for some anti-vaxxers on the internet, Charlie is not uniquely a man's name.

help eradicate the disease, licensing a vaccine within five years of the first clinical trials. The vaccine used in this case is called a viral vectored vaccine. These types of vaccines use a weakened virus as a shuttle that is engineered to deliver a genetic payload from the pathogen you are trying to prevent.

Five years sounds fast but it is nothing compared to the situation in which we found ourselves in 2020. The gene sequence of the virus that caused the COVID-19 pandemic, SARS-CoV-2, was published on 10 January, fifty-four days after the first recorded case. Sixty-three days later, on 13 March 2020, researchers from Moderna injected the first doses of a human vaccine. *Sixty-three days*. The word unprecedented gets thrown around a lot, but this is an unbelievable medical breakthrough. There are viruses on which we have been working for a vaccine for over sixty-three years, with no result. By September 2020, more than two hundred different vaccine candidates had been proposed (some more sensible than others) and thirty-six of them had made it into human clinical trials;[7] by November, three vaccines had demonstrated protective efficacy;* by the beginning of December, the first doses were being administered to people in the UK, with over 130,000 doses administered in a week; and by early February 2021, ten vaccines had been approved and more doses of COVID-19 vaccine had been administered than cases of COVID-19 infection recorded.

It is probably worth a quick interlude here to describe how vaccines are tested and ultimately licensed. The process has come a long way since the beginning. Pasteur and Jenner's experiments had a bit of a cavalier approach to

* Four if you include the Russian Sputnik vaccine.

them: it isn't recorded whether those involved had a real say in whether they wanted to be involved or not. Human experimentation pretty much stayed this way until after the Second World War when, following the terrible crimes of Dr Josef Mengele, the Nuremberg Code was adopted, protecting the rights of people in clinical trials. For vaccines (and other drugs), there are traditionally four phases of testing. Phases I–III happen before a product is licensed and phase IV trials monitor for unexpected reactions in a larger population after licensure.

The priority in these trials is safety, so before the vaccine is even tested in people it is tested in animals for acute toxicology – is it poisonous? Phase I (safety) clinical trials are conducted in a small number of healthy volunteers, often starting with a low dose and then escalating: they are sometimes called first-in-human studies. Phase II is basically a larger version of phase I, the principle being the more people something is tested on the greater the chance of detecting unexpected adverse effects. The immunogenicity will also be measured in the first two phases – does the vaccine lead to the production of an immune response? However, these early studies do not tell us whether the vaccine protects against infection. This can only be determined in phase III clinical trials; traditionally, randomised control trials (RCT). Randomised control trials are the gold standard for testing medical interventions. Researchers assign the volunteers to one of two groups: the experimental vaccine or the placebo control – an injection that looks exactly like the vaccine but doesn't contain any active ingredients. The important thing is that neither the participant nor the investigator who works with them knows who has had the vaccine; this is described as blinding. At the end of the trials the number of infected people in the vaccinated group is compared to the control group; this big reveal is known as unblinding.

The data is then reported as vaccine efficacy – presented as a percentage of the number of infections prevented:*

$$VE = \frac{\text{Infections in the unvaccinated group} - \text{Infections in the vaccinated group}}{\text{Infections in the unvaccinated group}} x100\%$$

Using actual data from the Pfizer/BioNTech COVID-19 vaccine trial,[8] there were 162 cases in the unvaccinated group and 8 cases in the vaccinated group. Assuming the two groups were the same size (more or less), then:

$$VE = \frac{162 - 8}{162} x100 = 95\%$$

Hence the reported 95% efficacy,† though this doesn't necessarily mean that the vaccine will protect 95% of people vaccinated in the actual world, but that under the conditions tested this is what was seen. There reflects a subtle difference between efficacy and effectiveness: efficacy is measured in a clinical trial – effectiveness in the real world. Hopefully, the trial is designed sufficiently robustly that it reflects the real world, but there may be multiple reasons why effectiveness is better or worse than trial efficacy.

Having demonstrated the vaccine is safe and protects against the pathogen, it then needs to be licensed by regulatory authorities. These are region specific: amongst others the MHRA covers the UK, the FDA covers the USA, the EMEA covers the EU and the WHO covers regions without their own specific agency. The purchase and roll-out of the vaccine is separate to the licensure and is down to bodies

* The second equation, as promised in Chapter 6.

† Told you it was worth waiting for.

that look at the cost per dose and the value per life saved; in the UK this function is performed by the JCVI. The regulatory authorities are politically independent and go on the science alone and nothing else.

All of this normally takes time. So where were the time savings made in 2020? As with all things COVID-19, the speed of innovation didn't come from nowhere; it stood on the shoulders of decades of research. Building on the experience of the Ebola outbreak, several organisations – for example CEPI (the Coalition for Epidemic Preparedness Innovations) – initiated programmes to prepare for a pandemic of an unknown pathogen – often referred to as Pathogen X. Vaccines were a central part of these preparations. CEPI funded cassette-style platforms, where the antigen could be swapped in from whatever nasty bug had cropped up.

There were many different cassette approaches proposed and tested. One platform that builds on the biotech revolution is DNA vaccination. In 1993, Jeffrey Ulmer at Merck demonstrated that you could generate an immune response in mice when you directly inject them with DNA rather than protein. There has been much interest in this approach since then. My team, especially Adam Walters, Ekaterina Kinnear and David Stirling, works closely with a company based in an old water treatment works in Kingston called Touchlight Genetics, to test the DNA vaccine platform.

Another type of genetic material has proved extremely (and surprisingly) effective in the race for a COVID-19 vaccine – RNA. The cellular machinery that translates RNA into proteins does not discriminate between RNA made in our cells, a virus or a factory in Norwood, Massachusetts. This means you can take RNA that encodes a viral protein, in this case the spike protein from SARS-CoV-2, inject it into someone's arm and the muscle cells will make the vaccine for you *in situ*. Since this approach can be applied to any protein from any organism, the platform can be rapidly adapted between different pathogens – hence the speed.

The first vaccine past the finishing post was an mRNA vaccine, developed by BioNTech/Pfizer. It's at this point that I feel obliged to point out that I knew BioNTech before they were famous – in fact Drs Laura Lambert and Charanjit Singh in my team published a study with them to show how RNA could be used as a vaccine against influenza back in 2017.[9] BioNTech used a subtly different platform in their final COVID-19 vaccine. One of the vital breakthroughs built on the work of Dr Katalin Karikó, who fled Communist Hungary in 1985 with $1,200 hidden in a teddy bear. Karikó realised that the way cells sensed RNA molecules could potentially be counterproductive for an RNA vaccine (see the section on PAMPs in Chapter 5). If the cell decides that the material is too foreign, it will not produce a protein from the mRNA – a vital flaw when you are trying to trick the body into making the vaccine for you. Karikó modified the mRNA in the vaccine to silence it and massively improved the potency of the vaccine. Thanks to her insight the resulting vaccine had 95% efficacy and roll-out began in mid-December in the UK and then in other countries. Brilliantly, one of the first people to get dosed was called William Shakespeare, showing us all's well that ends well.*

The team at Imperial College developed a variant on RNA vaccines. Professor Robin Shattock, with whom I have

* Other puns include: the one gentleman of corona, Vacbeth; 'Is this a needle which I see before me?'

shared a lab since 2008 (science name-drop siren), developed a self-amplifying RNA (saRNA) vaccine. This technology utilises one virus to defend against another. The saRNA vaccine is derived from a virus called an alphavirus and the RNA can make copies of itself in the cells it is injected into. Since it can replicate, the dose of vaccine can potentially be much smaller; you can vaccinate more people for the same amount of material. At the time of writing, thanks to epic efforts by Hannah Cheeseman, Kat Pollock, Jess O' Hara, Leon McFarlane, Paul McKay, Katie Flight, Tom Cole and the rest of Robin Shattock's team, the vaccine had gone from a standing start to being tested in humans in eight months.

The other vaccine to show rapid progress was ChAdOx1 nCoV-19. This vaccine was developed by the team at Oxford University (hence the Ox), led by Professor Dame Sarah Gilbert. It is similar to the vaccine licensed for Ebola, using a weakened virus to deliver a genetic cargo. The vector they use is derived from a chimpanzee adenovirus (hence the ChAd). The core insert of the vaccine – the antigen that protects us from SARS – was co-designed over the course of a weekend by Professor Teresa Lambe OBE. Speaking to Tess in a snatched WhatsApp chat between her clinical trials, she told me that: 'It was a crazy time, it's been a crazy year, but seeing our vaccine come to fruition has been one of the most rewarding experiences of my scientific career, though I'd welcome a quieter 2021.'

Working with AstraZeneca, this vaccine moved rapidly into large-scale clinical trials in the middle of 2020. Professor Sir Andy Pollard coordinated the trials, building on his many years of experience in clinical trials. He reflected on the unique situation in 2020:

> During the pandemic, we have been inundated with interest from volunteers who have been prepared to take part in clinical trials to see if our vaccine could control the virus, and international researchers have made themselves available at short notice to work tirelessly on our shared goal to develop a not-for-profit vaccine for all corners of the world. Despite the goodwill, none of the teams were prepared for the extraordinary efforts that have been required to manage pandemic logistics, and the resilience needed to keep going through lockdowns, illness among the trial teams, and the enormous and sustained workload for the past 11 months, all with the whole world watching our every move. As we come to the final steps, we hope that our vaccine will have an impact that saves lives, reduce the pressure on health systems, and contribute to the return of the normal human interaction for which we all yearn.

Notably, the Oxford/Astra vaccine was made and distributed at cost for the benefit of everyone. It was rapidly licensed to other companies in order to distribute the manufacturing worldwide; particularly the Serum Institute of India, who anticipated making one billion doses in a year. Following approval by the WHO in February 2021, the vaccine was rolled out across the world.

There are several books' worth of comment on the COVID-19 vaccines, so I will summarise by saying it was incredibly fast and by the time you read this we will know whether it worked or led to a zombie apocalypse – presumably the former or you wouldn't be reading this.

INFECTION

But the coronavirus and Ebola virus vaccines aren't the only new kids on the block. Vaccines for both malaria and typhoid have been licensed in the last five years. They use different platforms to each other, but they have both built on data from controlled human infection models; where volunteers are deliberately infected with a pathogen. Human challenge has had a long history in vaccine development, right back to Edward Jenner. It moved to a more modern framework in the 1970s, under the auspices of Myron Levine in the US; initially in prisons but later in a hospital. Levine was interested in cholera. Cholera causes an unpleasant disease and yet he still found volunteers to be infected – one of whom required twenty-six litres of intravenous fluids to replace what he had pooped out.[10] Levine's challenge model stimulated further research, leading to the establishment of a typhoid model by Andy Pollard at Oxford (who also led the clinical trials for the Oxford/Astra COVID-19 vaccine). This model led to the licensing of a new typhoid vaccine in 2018, replacing Almroth Wright's inactivated one.

Around the same time that Levine was getting people to crap voluminously, the Common Cold Research Unit on Salisbury Plain, UK was investigating exactly that: following what happened when researchers infected volunteers with common colds. This led to incredibly useful data that we still use today. For example, investigators discovered an antibody threshold above which you should be protected from influenza. Knowing the protective threshold makes developing vaccines considerably easier. David Tyrrell first discovered coronaviruses at the Common Cold Unit in collaboration with June Almeida, an electron microscopist working at St Thomas' Hospital.[11] From their discovery in 1966, coronaviruses remained a reasonably unloved corner of viral research until January 2020, when everyone and their dog suddenly became an expert on them.

The malaria vaccine called RTS-S, but marketed as Mosquirix, also drew on data from controlled human infection studies. While viruses and bacteria for infection can be grown relatively easily and then squirted up the nose (for influenza) or into the stomach (for typhoid), malaria is vector-borne. It must be delivered by mosquito bite. This means you need to keep the malaria parasite alive and maintain mosquitoes in an insectary. Mostly mosquitoes eat nectar, but the female mosquito (from some species) needs a blood meal to acquire enough iron to lay eggs: male mosquitoes never bite. In the lab, mosquitoes can be tricked into biting through a rubber glove into a jar of blood. To perform the human challenge studies, the mosquitoes are fed some infected blood and then put into a jar Sellotaped onto the arm of the volunteer, until they are bitten five times. From which we can conclude that volunteers for human challenge studies are a special breed: not only do they tolerate getting an infection that might necessitate twenty-six litres of intravenous fluid, but they will also tolerate getting five mosquito bites. I know which I'd prefer.

PREDICTION

So here we are in 2021. The question is, what's next? The big three (HIV, TB and malaria) still need vaccines that work. Yes, vaccines exist for malaria and TB, but they are not completely effective. Mosquirix, the first malaria vaccine, is only about 40% effective in the first year after vaccination and BCG is only about 30% effective in adults. An HIV vaccine has long been a research goal, but it is proving to be incredibly challenging. Other vaccine targets include a universal flu vaccine – one that covers all of the different variants – and vaccines against the sexually transmitted infections *Chlamydia trachomatis*, gonorrhoea and syphilis. There also remains a whole slew of neglected tropical diseases without vaccines including chikungunya, hookworm, leishmaniasis and Chagas disease.

Entirely speculatively, I predict the next two vaccines to be licensed will be against RSV and TB. From a personal point of view, an RSV vaccine will be a mixed blessing: it will be wonderful to see an end to a disease I have spent many years working on, but it will certainly reduce my ability to get further research funding for it in the future.

I think an RSV vaccine is imminent because of the breakthrough research by Peter Kwong, Barney Graham and Jason McLellan at the NIH. Normally viral proteins can change conformation 'like a Transformer toy', in Graham's words.[12] The NIH team demonstrated that you could genetically stabilise the key viral antigen from RSV (the F protein) to make it much more protective. This information proved vital in the development of the vaccines for COVID-19.

The new RSV vaccine might be given to pregnant mothers as a way to protect their babies. This is a variant of the injection of horse serum in the 1920s to provide passive immunity. Mothers pass on immunity to their children both during pregnancy, via the placenta, and after birth in the milk. Working with Professor Beate Kampmann at the London School of Hygiene and Tropical Medicine, Miko Zhong in my group investigated how the timing of maternal immunisation affected the amount of antibody transferred to the baby.[13] Protecting children by vaccinating mums is already used for influenza and tetanus and in the future it may well be used for group B streptococcus (GBS) and RSV.

A new TB vaccine would be an even bigger story. TB is a huge global killer, hitchhiking on the back of the HIV pandemic. In 2000, two million people died of TB world-wide. TB deaths haven't declined much in the last twenty years – 1.5 million people still died of TB in 2019. GSK has developed a vaccine based on two mycobacterial antigens (called MTB32A and MTB39A), but these antigens do not work on their own; they need to be combined with an additional magic ingredient called an adjuvant.

Adjuvants increase the potency of an administered vaccine. In 1924, Gaston Ramon, a French vet, first noticed that the amount of antibody produced increased when the horses developed an abscess.[14] Ramon has the unfortunate record of being the person to receive the most Nobel nominations without ever winning the prize – having been nominated 155 times between 1930 and 1953. Then again, he also tested breadcrumbs and tapioca to improve vaccine responses, so maybe he was less genius and more tinkerer. Recent studies suggest he wasn't completely missing the mark, because Dr Ryan Russell in my lab showed that tiny grains of sand (Nano-SiO_2) could increase the potency of flu vaccines in baby mice.[15]

A range of compounds have been tested since unlucky Gaston first proposed adjuvants. GSK have developed a particularly potent adjuvant called AS01E (AS standing for adjuvant series). AS01E comprises oil droplets containing two active ingredients: 3-O-desacyl-4′-monophosphoryl lipid A (MPL) and QS-21. MPL is a derivative of LPS, which in turn is part of the cell wall of certain bacteria and, more importantly, one of the chemical signatures by which the immune system recognises that it is under attack (the danger signals identified by Charlie Janeway in Chapter 5). The inclusion of MPL makes sense immunologically. MPL activates the immune system and therefore improves the response to a vaccine. QS-21 is a bit more of a mystery. It is a purified plant extract that comes only from the soap bark tree (*Quillaja saponaria*), which is found exclusively in Chile. It definitely has an effect, by improving the immune response to vaccines, but the honest truth is we don't really know how. The inclusion of AS01E in the new TB vaccine peps it up, making the resultant vaccine (called M72) extremely effective: it reduced TB vaccine cases by 50% in early clinical trials.

Interlude

THIS IS ADMITTEDLY a long chapter.

That's because vaccines are Amazing.

But time for a quick pause for you to stretch your legs and maybe get a cup of tea.

Here's a fun fact for you to consider while the kettle boils.

> In the approximately thirty minutes it would have taken you to read the first half of the chapter, 5,000 children will have been vaccinated around the world, saving between 100 and 200 lives.

But I imagine you are now thinking that the history of vaccines is fascinating, but how do they work and are they really as effective as he keeps saying?

Well, take out the teabag, add the milk and read on.*

* And sugar, if you insist, you devil.

CHAPTER 9

How vaccines work and their incredible impact

Two fields of science underpin vaccines: immunology and epidemiology. Vaccines are based around the principle that the immune system can remember things it has seen previously. In the days before vaccines, this meant that if you had been infected with something and survived you were unlikely to get that same infection again. Immunisation manipulates this system, producing a memory to an infection you are yet to see, without any of the bad bits of the infection itself. Having seen the vaccine, your B cells produce antibodies specific to the pathogen. This means that if you then come across the pathogen your blood is chock-full of protective chemicals that can recognise and kill it. Vaccination also gives you a small, highly focused army of T cells primed to hunt down and kill any infected cell (see figure 9).

The immune system works at an individual level. But vaccines are most effective as a population intervention, the fundamental principle being to reduce R_0, or the number of people that any one infected person will spread their disease onto: if R_0 is less than 1 the epidemic contracts; if R_0 is greater than 1 it spreads. Altering the dynamics of an infection with vaccines means you can protect the whole population without immunising the whole population. This population-level immunity is called herd immunity. Herd immunity even protects people who don't respond as well to the vaccine, for example the immunocompromised.

The percentage of the population that needs to be immune to achieve herd immunity varies – the more infectious a pathogen is, the higher the vaccination rate needs to be. For a pathogen with a low R_0, such as *Haemophilus influenzae* B, you can reach herd immunity when only about 25% of the population are immune, but for pathogens with a high R_0 like measles you need to immunise upwards of 90%. When applied properly, vaccine-induced herd immunity can be a potent tool. The vaccine against *S. pneumoniae* (sometimes called pneumococcus) is given to infants but provides two in one protection: the child doesn't get sick and it does not pass its germs on to its grandparents.

Epidemiological understanding of how pathogens spread also means vaccines can be focused on the correct populations. This builds on the point made in Chapter 3 that if you are never exposed to a pathogen then you don't need a vaccine against it. In a previous less-woke age, I would illustrate this with the Slappers and Nuns model, but now I probably should call it targeted vaccination – but the principle remains the same. A vaccine against sexually transmitted infections will be more effective in people who are sexually active compared to those who have taken vows of celibacy.

The population targeted for vaccination can be decided by the disease caused as well as the risk behaviour undertaken. In the UK the HPV vaccine was initially only given to girls, because the vaccine predominantly prevents cervical cancer, which is much harder to get if you don't have a cervix. It is now given to both boys and girls because it also

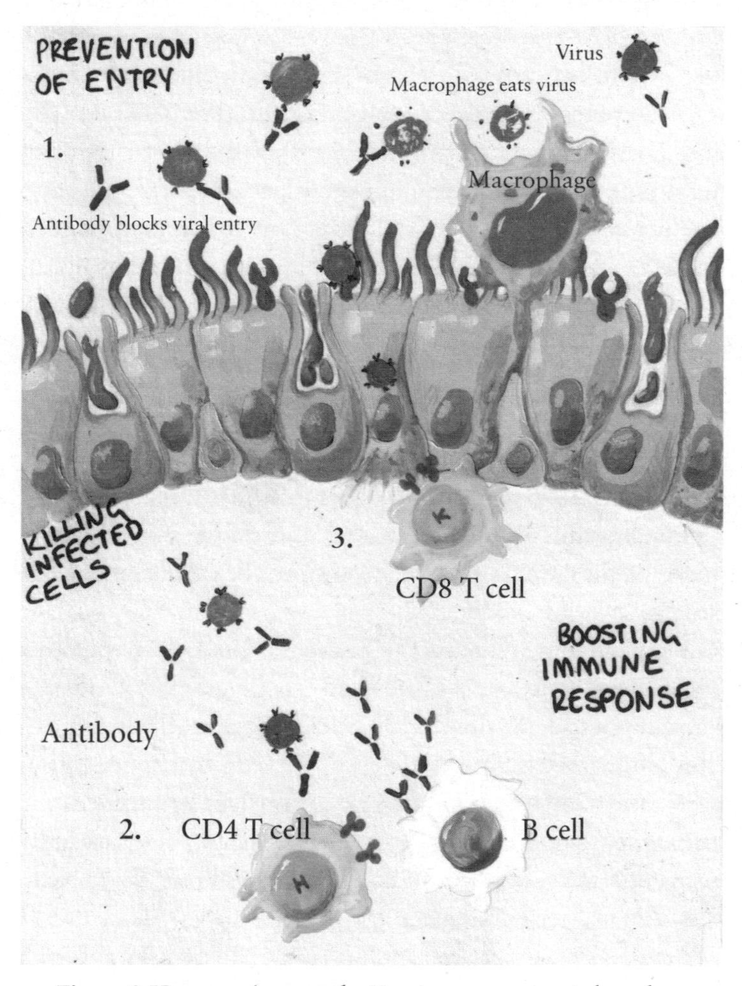

Figure 9 How vaccines work: Vaccine protection is based on immune memory mediated by T and B cells. 1. Antibodies prevent infection – they block the virus from getting into the cells. 2. The CD4 T cells boost the quality of the response. 3. CD8 T cells kill infected cells.

prevents genital warts that can cause testicular cancer, infected boys can still spread the virus and the herd immunity benefit was not reaching into the MSM (men who have sex with men) community when only girls were vaccinated. The HPV campaign in the UK is one of the stories that highlights the remarkable power of vaccination. A very recent study showed a 90% reduction in cervical cancer since the introduction of the HPV vaccine: ten times fewer people got this type of cancer.[16]

CELEBRATION

And yet, despite the clear positive impact of vaccines, we live in an age, especially in richer countries, when people remain hesitant about their use. Before addressing why this might be the case, I think we need to celebrate the incredible success story of vaccination.

I always come back to the numbers. Let's start with the UK: in 1940 there were more than 60,000 cases of diphtheria, leading to 3,283 deaths. By 2002, vaccination had almost eliminated it – with just two deaths from diphtheria between 1986 and 2002. Meningitis C, invasive pneumococcal disease, *Haemophilus influenzae* type B and rotavirus have been virtually eradicated. Polio and smallpox have both been completely eradicated.

These successes are reflected on a global level. The WHO estimates that vaccines against diphtheria, tetanus, whooping cough and measles prevent 2–3 million deaths every year. In 1963, when the vaccine was introduced, measles killed 2.6 million people a year; this has been cut globally to fewer than 100,000 – which is still far too many. Likewise, childhood deaths from tetanus reduced from 800,000 in 1988 to 50,000 in 2013. This reduction of vaccine-preventable disease in the last twenty years is down to the efforts of multinational organisations such as GAVI (Global Alliance for Vaccines and Immunization) and support by governments and philanthropists such as Bill and Melinda Gates. The percentage of children under one who got at least one vaccination was 22% in 1980; it had increased to 88% by 2016.

If these everyday wins are not enough to persuade you, what about the big-ticket items? Let's start with polio. In 1981 there were about 460,000 reported cases of polio worldwide and in 2019 there were 545 cases, a thousandfold reduction. In August 2020, bucking the trend for what was a pretty sorry year, Africa was declared free of wild type polio, meaning five out of six global regions are free of this disease. Only two countries now have wild type polio virus circulating: Afghanistan and Pakistan. If you compare the number of cases vs the world population (and admittedly this is not a globally spread risk) the chances of catching polio are one in two million, the same as tossing twenty-one heads in a row: in the USA you are considerably more likely to be injured on the toilet (22.5 cases per 100,000 population or a 1 in 5,000 chance).[17]

There is a slight complication to this story. Africa is not completely free because live vaccine that has been pooped out can be infectious. Unfortunately, Sabin's vaccine was not quite as stable as he thought. It is perfectly safe in the vaccinated person, but if left to its own devices in untreated sewage the virus can revert to a more virulent type – viruses being tricky like that. There are now more cases of secondary infection from contaminated water than wild type virus – but the numbers are vanishingly small for both. In 2019 there were 369 cases of vaccine strain-derived disease, compared to 176 of wild type virus. One of the pieces of good news that may have been overlooked in the deluge of misery that was 2020 was the emergency licensure in November of a new live attenuated polio vaccine. This new vaccine – nOPV2, made by Bio Pharma in Indonesia – has been genetically engineered to be more stable and will be deployed to prevent vaccine-derived polio infection.[18] Hopefully this is the beginning of polio's end.

Probably the biggest triumph of vaccines is the complete eradication of a pathogen. The best-known example is smallpox, rightly celebrated as a triumph of modern medicine. In 1959, the World Health Organization (WHO) started a campaign to eradicate the virus. The first attempt failed, mainly due to a lack of funding. A second intensified programme was launched in 1967, leading to the eradication of smallpox in South America in 1971 and Asia in 1975, leaving only Africa with cases of the virus. In 1977 Ali Maow Maalin, a Somali cook, was the last person to be naturally infected with smallpox. While he survived, the tale had a tragic twist. In 1978, Janet Parker, a medical photographer working at the University of Birmingham, contracted smallpox and died. How she came into contact with the virus is not clear; the lab on the floor below hers worked on the virus but smallpox was rarely transmitted through the air. The tragic death of Parker led to the global destruction of all stocks of the virus except for those at the Centers for Disease Control (CDC) in the USA and VECTOR, the alarmingly named equivalent in Russia (previously a hub for biological weapons research). Since 1978, there have been no cases of smallpox anywhere.

The other, sometimes overlooked, virus to be eradicated is rinderpest. Rinderpest is related to the measles virus and was very bad news for cows, killing nearly 100% of infected animals. It is another of Genghis Khan's legacies, as it followed his hordes in the baggage trains.* In the 1960s, Walter Plowright developed an attenuated vaccine against rinderpest, using a similar approach to Sabin of multiple passage through cells. This highly potent vaccine broke down at higher temperatures, which was a problem for getting the vaccine to where it was needed most – cattle herds in sub-Saharan Africa. An improved version of the vaccine, stable for two weeks at 45°C, was then developed. Working closely with farming communities the Food and Agriculture Organization (the WHO for animals) targeted vaccination programmes, leading to the eradication of the virus by 2010. The rinderpest vaccine is just one of a multitude of vaccines that are available for all kinds of animal species. There are a lot of animal vaccines out there, including those for cats, dogs, pigs and even fish. Thanks to animal vaccines, Fluffy, the ageing tomcat, can live long enough to eat salmon for tea.

THE STATE WE'RE IN

But vaccines only work if you take them. Which is kind of obvious. If you want to see what happens when people stop getting vaccinated, look no further than the reappearance of measles. Measles, caused by the measles morbillivirus, is particularly pernicious. It causes a severe disease in and of its own right, with a 10% mortality rate – 10 times higher than COVID-19 and 100 times higher than flu. If you survive, measles also screws you up long term. Measles increases the

* It has been suggested that 0.5% of all men worldwide are direct descendants of Genghis Khan.

risk of death by any other pathogen up to a year after you recover from the original infection. It does this in a subtly unpleasant way: it infects your B cells (the ones that make antibody) and kills them, erasing all those protective memories.

Measles combines two problems as a re-emerging pathogen; one biological and one sociological. The biological problem is that measles has an extremely high R_0 value – in an unvaccinated population, one person infected with measles can potentially infect up to eighteen other people. This problem can multiply, because people who choose not to vaccinate often cluster together socially. For example, in June 2005 an unvaccinated volunteer from Indiana went to Romania to work in an orphanage and became infected. On returning to their church (where fifty out of five hundred members were unvaccinated) to talk about their work they managed to infect sixteen people at a single gathering.[19] The outbreak led to thirty-four cases and cost about $167,685 to contain, compared to the $50 it would have cost to immunise the rest of the congregation.

The sociological problem is the skirling mass of misinformation about the vaccine. This comes down to one man, ~~Dr~~ Andrew Wakefield. In 1998 Wakefield published a paper in the medical journal *The Lancet*. In the deeply flawed, widely discredited and finally retracted paper, Wakefield claimed a link between MMR and autism. A generous interpretation of events is that Wakefield, desperate for academic success and publicity, made a mistake and then dug himself too big a hole to escape.* However, a newspaper reported

* If he was seeking fame, he has certainly achieved this goal – though personally I would prefer to remain unknown rather than accept the cost of Wakefield's fame.

that Wakefield had received a payment of £55,000 from solicitors seeking evidence against vaccine manufacturers, data in the paper was manipulated and several of the parents involved in the study were litigants. Wakefield was struck off as a doctor in 2010. There is no link between MMR and autism. The evidence that MMR does not cause autism is overwhelming; the link has been disproved by multiple studies (twelve so far) including one involving 1.8 million children in Finland. On some levels this should be enough to convince people – it's certainly enough to convince me. But sadly, some weeds have deep roots.

So why don't people get themselves or their children vaccinated? The reasons are complex and vary from country to country.[20] They mostly come down to people making a cost–benefit calculation using skewed data. Vaccines do, of course, come with a cost: both in terms of the time taken going to the clinic and in cash, depending upon the healthcare setting. Vaccines also come with a level of risk. They are acutely painful, with most injections leading to some pain at the injection site and a smaller proportion leading to headaches and a slight rise in temperature. When given to babies this can lead to sleepless nights, which may deter parents from getting a second dose.

There has been much research on vaccine-induced inflammation. Dr Jacqueline McDonald in my lab looked at how age changes responses to vaccines.[21] The inflammation comes as a necessary side effect to how vaccines work: the immune reaction to foreign material will cause localised swelling and pain. Vaccination is a trade-off – sore arm now versus death later! There is also a much, much smaller risk of long-term effects from vaccination. These risks are vanishingly small – for example, there have been thirty-five deaths from yellow fever vaccine since 1935, which is awful, but the denominator is more than 400 million; so a 0.000007% chance, compared to a 15% chance that yellow fever will kill you if contracted. Let's use the jelly bean analogy to illustrate this – would you take a jelly bean from a bowl if you knew that one in every 10 beans would kill you (the virus) or from a bowl where one in 1.5 million beans might kill you (the vaccine).

These costs and risks come up front and can be highly publicised – every yellow fever vaccine death will receive global news coverage, while every yellow fever death goes unremarked, except by the families affected. And therein lies the problem. Vaccines reduce illness from infectious diseases, dramatically. Vaccine hesitancy tends to be greater in richer countries. When the likelihood of knowing someone with a vaccine-preventable illness is very low, the incentive to vaccinate reduces. In this respect vaccines are very different from a cancer drug. Nearly everyone will know someone who has died, unpleasantly, of cancer. Cancer drugs cause pretty grim side effects: chemo makes your hair fall out and the new immunotherapy drugs can lead to chronic gastroenteritis (they make you poo yourself). But people take them because they are preferable to the alternative – certain death.

Humanity's hard-wired inability to understand risk is clearly demonstrated by the desire to do things that are clearly and indubitably bad for you: smoking, drinking, going to roller discos. The present us is very bad at predicting the impact of our actions on the future us. We will happily trade a better now for a worse future and then complain when the future becomes the now. Richard Thaler in *Nudge* describes ways to enable subtle changes in human behaviour, noting that while Econs (theoretical beings that are perfect at weighing up strategies) might make the right decisions, Humans rarely do.[22] I will return to conspiracy theories, misunderstanding and misinformation in Chapter 15.

A further issue with vaccines is that they aren't only an individual medicine: they have a societal benefit. One of their benefits is to induce herd immunity; meaning that the most vulnerable never find themselves exposed to the pathogens. By getting yourself or your family vaccinated you are doing a civic duty and protecting others. I need to weigh this up each year when I get my influenza vaccine. I know the risk of severe influenza to me is low, but the risk that I catch it and pass it on to my parents is much higher. Vaccines form part of the social contract – the same as wearing masks, social distancing and washing hands were during COVID-19.

What can be done? Immunology circles spend a lot of time hand-wringing about anti-vax, but I don't think it is the worst thing going on in the world right now (global warming, antibiotic resistance and vegan chicken nuggets vie for that spot). People not liking vaccines isn't a new problem. While there was much enthusiastic uptake of Jenner's vaccine programme, there were also anti-vaxxers from the start – a cartoon published in 1802 showed vaccinated people sprouting cows from their ears. However, most people aren't anti-vax; instead, they are more vaccine unsure and as such it is better to call it vaccine hesitancy than anti-vax. Telling someone that they are wrong because they couldn't find time to go to a vaccine appointment will not encourage them to get another appointment. One simple solution is to improve access to vaccines. Sometimes children go unvaccinated because there is no vaccine at the clinic or there is not the right member of staff to administer it at the time of the appointment.

Another approach that can improve uptake is the application of behavioural economics, or nudges.[22] These make it easier for us non-rational humans to behave more like Econs and do the right thing. One suggestion is to make vaccination the default option – meaning you have to opt out rather than opt in; this simple step means that parents aren't having to remember to fill in a form to get their children vaccinated. The simpler the process is, the more likely people are to choose it.[23] Nudges can be extremely effective: the sugar tax on fizzy drinks in the UK has led to a 10% reduction in the sugar consumed.

I think it is important not to over-promote the anti-vaxxers; they are highly vocal, but are a very low proportion of the population. In a 2018 study on attitudes to vaccines, which covered most countries in the world, the majority of people either strongly agreed that vaccines were safe or were not sure. The numbers who thought vaccines were unsafe essentially tracked at 0%.[24] And there is a big difference in saying you are a bit unsure and not getting the vaccine.

Access to the right information in a way in which it can be understood and not misinterpreted is vital. The global polio vaccination campaign has all the ingredients of a ripe conspiracy theory. It has examples of error (the Cutter incident), unexpected complications (occasionally infectious faeces after oral polio vaccination), international players (the WHO and the Gates Foundation), international intrigue (the CIA set up a fake vaccine campaign to find Bin Laden) and fake news (rumours circulated in Nigeria linking vaccination to sterilisation, leading to a widespread boycott). As with all interfaces between medicine and politics, it is a complex situation that can easily be misdescribed in a tweet or a Facebook post.

Another example is the story about mercury in vaccines. The preservative compound called thimerosal contains a compound of mercury (ethyl mercury) and thimerosal was used in multi-dose vaccines to prevent contamination, so people were not given an infection at the same time as their injection; which I think we can all agree is a good thing. There is no evidence that thimerosal was unsafe; a Danish study covering three million children found no evidence of a link between thimerosal and autism.[25] Even more relevantly, it hasn't been used in vaccines since 1999. But finding all of that information takes time and effort and mercury sounds nasty and mercury poisoning, thanks to the Mad Hatter, is in the collective consciousness so it is easy to jump to conclusions, especially if swayed by 'influencers'. Organisations such as the NHS and CDC have gone a long way to make the information easier to find. I personally do not favour laws stipulating compulsory vaccination. They can increase resistance to the vaccines if they are seen as mandated and they can also have a bigger impact on disadvantaged groups who may not be able to easily access vaccines.

In 2020, we got to watch all of this play out in real time. Vaccines comprised part of the strategy to control the COVID-19 pandemic. They were developed at a far greater speed than ever before. This was achieved through massive investment, both in the research to generate the new candidates and in the manufacturing. To coordinate the delivery of the vaccine with the end of the clinical trials, doses of vaccines were made before it was known if they were safe or effective, some of which will no doubt end up going down the drain. Reassuringly, the clinical trials were performed to the same high level of scrutiny as before. For example, the AstraZeneca trial which enrolled 23,848 people was paused in September 2020 because of adverse effects in *one* individual. The manufacturers of the vaccines also signed an accord that they would not release the vaccines until they were properly scrutinised by regulators. This was all against the background of increased global uncertainty, which can raise the profile of conspiracy theories. In spite of all this, large numbers of people received the COVID-19 vaccine in the UK, with 93% uptake in over 75-year-olds. In areas with lower vaccine uptake, the main drivers were accessibility issues and a preference to use local GP surgeries run by familiar staff, not centralised centres run by strangers; anti-vax sentiment had very little impact.

In the end, vaccines are a vital tool in the prevention of infection. They are a success story of human science and innovation, up there with the moon landing, Swiss Army knives and the internet. One wonders what Jenner would think about the process he had started. Given he was an eighteenth-century GP working in a farming community, he would probably have poured a bumper of port and got sloshed with his friends – not a bad response in my opinion.

However, vaccines are not the whole answer to infection and despite our best preventative strategies, sometimes we do get infected and need treatment. Luckily, drugs exist that can cure infections: bacterial, viral, fungal and parasitic. I will start by exploring antibiotics – drugs that treat bacterial infections, for example penicillin. The next chapter will discuss how the drugs work and the fight back by bacteria. It will explore the people who invented them, why the antibiotic pipeline is drying up and what is being done to restart it. It will also come back to the microbiome and how our relationship with our bacteria can be disrupted.

CHAPTER 10

Antibiotic

Timeline: Early October 2020. Epsom. Overly complicated tier system implemented in UK. Global COVID-19 cases 34,495,176; deaths 1,025,729.

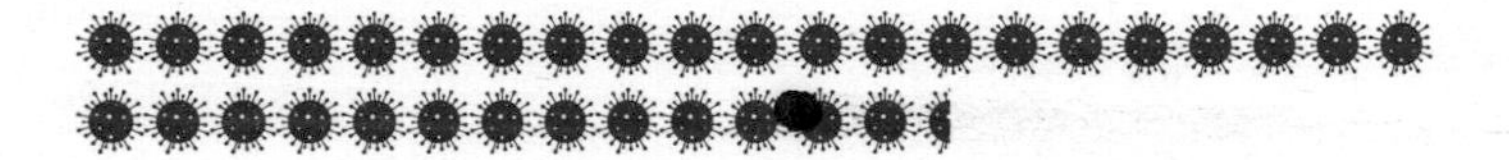

'The trouble with being a hypochondriac these days is that antibiotics have cured all the good diseases'

Caskie Stinnett

ON 30 JUNE 1924, Calvin Coolidge Jr (son of the twenty-ninth president of the USA) made the fateful decision to play tennis without socks. He developed a blister on one of his toes, which turned septic and killed him seven days later. This raises several questions: how dirty were his tennis shoes and had he perhaps considered washing his feet after the game? But more importantly it emphasises the risks of life in the pre-antibiotic age. We have, understandably, become more focused on viruses in 2020, but bacteria are a considerable and resurging threat to human health. Had Calvin Jr kept his socks on till 1928 things might have turned out differently, because on 28 September 1928

Alexander Fleming was about to, in his own modest words, 'revolutionize all medicine by discovering the world's first antibiotic' (see figure 10).

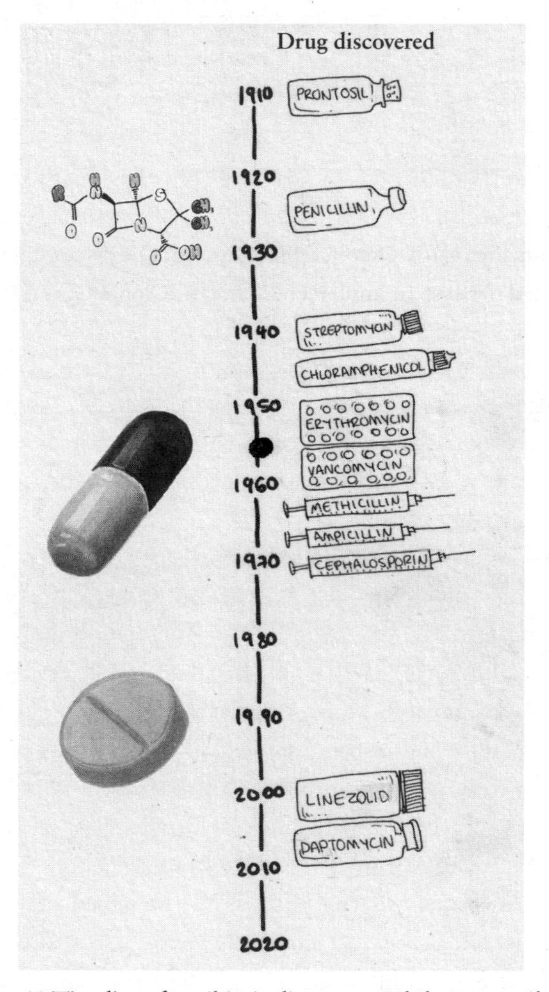

Figure 10 Timeline of antibiotic discovery: While **Prontosil** was the first antibiotic, the discovery of **penicillin** and its mass-scale production in the 1940s led to a surge in antibiotic discovery and development.

A BRIEF HISTORY OF ANTIBIOTICS

But the history of antibiotics doesn't begin with Fleming and penicillin. The ancient Egyptians applied mouldy bread to wounds. Whether or not they knew why this worked is unclear, but they also invented beer, so we should give them some credit for being the first microbiologists. Evidence for the ancient Egyptians as early antibiotic adopters comes from archaeological research. The tetracyclines (a type of antibiotic) bind to bones and have been detected in skeletons from Nubia dating from 350 to 550 CE. Another pre-antibiotic strategy was to apply maggots to wounded tissue. Remarkably, this works: the maggots preferentially eat the dead tissue. Even more remarkably, maggot therapy is making a comeback for hard-to-treat wounds such as diabetic ulcers: available at all good pharmacists and tackle shops near you.

Our old friend and scientific duellist Félix d'Hérelle took a more rational approach to the prevention of bacterial infection. Having discovered the bacteriophage (viruses that infect bacteria) in 1917, by 1919 d'Hérelle had successfully isolated enough phage from chicken poo to cure someone of dysentery. In the absence of other therapies, this approach took off, particularly in the USSR – which led to d'Hérelle's run-in with the NKVD. However, phage therapy was unreliable, with inconsistency between batches leading to variable efficacy.[1] Drugs ultimately replaced phages – as simple chemicals they are easier to synthesise and check for quality control. However, phage therapy did continue, particularly in the USSR, due to the isolation from Western scientists.

ZAUBERKUGEL

Paul Ehrlich developed the first deliberate antibacterial drug in 1909; yes, the same Paul Ehrlich who discovered antibodies also developed one of the first methods for staining bacteria and won a Nobel Prize. Which makes me reflect whether science was just easier in the late nineteenth and early twentieth centuries. Pasteur not only proved the existence of bacteria but he also produced the first bacterial vaccine; Marie Curie discovered radioactivity, radium *and* polonium; and Ehrlich founded two therapeutic fields – antibodies and antibiotics. I like to think that there were either simpler questions to be answered or fewer scientists. This helps me square the fact that I have been working for twenty years and have discovered that science is quite hard and most of my hypotheses turn out to be wrong.*

Back to Dr Paul smarty-pants Ehrlich who, when he wasn't busy curing diphtheria by injecting horses, had observed that some bacteria took up dyes more readily than others. Building upon this observation, he theorised the existence of toxins that could selectively kill bacterial cells but not human ones. He called these magic bullets, or *Zauberkugel* in German. German is simply better for compound words; see also *Schadenfreude* (joy-harm; laughing at others' misfortune), *Backpfeife* (slap-face; a face you would like to slap, e.g. Michael Gove's), *Torschlusspanik* (gate-shut panic; worrying about not hitting life milestones in time, like marriage or finishing a book edit), *Kummerspeck*

* In spite of all he did, Ehrlich's is not a household name – I suspect you would do pretty well in the 'Germans beginning with E' round on *Pointless* if you named him.

(sorrow bacon; weight gained from comfort eating) and *Nacktschnecke* (naked-snail or slug). COVID-19 provided ample opportunities to come up with more compound words, including *coronamüde* (tired of COVID), *Impfneid* (envy of those vaccinated) and *Hamsterkauf* (hamster-buy; the act of buying too much loo roll).

Sahachiro Hata in Ehrlich's lab screened hundreds of compounds (chemicals not words – though brilliantly the German word for drug screening is *Rasteruntersuchung*, itself a compound word). The six hundred and sixth compound, imaginatively called Compound 606, effectively killed spiral-shaped bacteria, including syphilis, caused by *Treponema pallidum* subspecies *pallidum*. Christopher Columbus's sailors brought syphilis back from the Americas alongside maize and chilli peppers: they'd clearly had quite a time of it. Syphilis reached Naples in the late fifteenth century, when French soldiers invaded, leading it to be called the 'French disease'. It initially presents as a chancre, basically a big ulcer, on the genitals, but does eventually spread to the brain. The madness of George III was attributed to syphilis, but was more likely to be porphyria, a genetic disorder. Syphilis killed a whole slew of artists including Toulouse-Lautrec, Gauguin and Manet, which suggests they did more than paint the prostitutes. Prior to Ehrlich and Compound 606, syphilis was treated with mercury. Presumably people would rather take mercury than the alternative, the dreaded urethral umbrella, which is as terrible as it sounds – a small device was inserted up the urethra and then opened up to clear blockages. In 1910 commercial production of Compound 606 began under the name Salvarsan. Salvarsan was a bit of a bugger to work with, being extremely sensitive to air and somewhat toxic.

Controversy surrounded Salvarsan, foreshadowing the furore around other interventions to prevent STIs, with some groups claiming it would lead to the breakdown of moral values.

The screening approach of Ehrlich and Hata set a precedent for future antibiotic discovery. Josef Klarer and Fritz Mietzsch took a similar line to Ehrlich by researching the antibacterial properties of dyes, while working at the German pharmaceutical company Bayer in the 1930s. In 1932, after extensive rounds of testing, they synthesised and tested a red dye with potent antibacterial properties, called prontosil-rubrum. They collaborated with a medical doctor, Gerhard Domagk, who was awarded the Nobel Prize for his role in the work. But rather than let him receive his prize the Gestapo arrested him and forced him to turn it down, because the Nobel organisation had angered the Nazis by awarding a Peace Prize to an anti-Nazi, Carl von Ossietzky. Domagk did ultimately receive the award in 1947. He wasn't the only German scientist to have Nobel Prize issues with the Nazis: James Franck (a Jew) and Max von Laue (a Jewish sympathiser) sent their medals to Niels Bohr in Denmark for safekeeping. Sadly, the Germans invaded Denmark. Rather than let the medals be captured, George de Hevesy in Bohr's lab dissolved their medals in acid and left them on a shelf. After the war, de Hevesy recovered the gold and sent it to Sweden to be crafted back into medals.

Bayer marketed Domagk's antibacterial drug as Prontosil. It received a welcome publicity boost in 1936 when another president's son, Franklin Roosevelt Jr, had a sore throat. Naming children after yourself is apparently a pre-requisite for the presidency: twenty-three out of forty-six US presidents named their children after themselves, or twenty-four if you count Lynda Johnson (daughter of Lyndon).* Unlike the hapless Coolidge Jr, there was now a treatment and Roosevelt Jr quickly recovered. This boosted profile of Prontosil was both good and bad for Bayer. Unfortunately for them the chemical they had patented (Sulfamidochrysoïdine) wasn't the active form of the drug. Prontosil is a prodrug that is metabolised in the body into the active form, sulphanilamide, which had been discovered twenty years earlier and was therefore off patent. This allowed other companies to get in on the action and make their own versions of the same drug, called the **sulphonamides** or 'sulfa drugs'. Generic sulfa drugs not only saved the life of Roosevelt's son but they also cured Winston Churchill (and Nero the royal lion) from life-threatening pneumonia in the winter of 1943. Sulfa drugs remained widely in use until penicillin replaced them but are still used today to treat infections with bacteria resistant to other antibiotics.

PENICILLIN AND THE LUCKY MOULD

The story of Fleming and his contaminated plates has often been retold. Basically, his lab was a bit of a dive; he hadn't tidied up before going on his summer holiday and he got contamination on his bacterial growth studies. For some reason best known to him, rather than saying 'oh no, I've ruined my experiment', he said, 'that's curious', and isolated the mould juice that could kill bacteria. The story goes on to tell how the mould spores came from the Fountains

* The 46th president, Joe Biden, is, like ten other presidents, named after his father.

Abbey pub across the road; and having had a pint there under previous owners I can confirm that was indeed a possibility. But having a dirty lab alone doesn't, unfortunately, win you Nobel Prizes – if it did, I could easily rank my PhD students in order of likelihood of getting a Nobel Prize. As with Pasteur's accidental discovery of the *Pasteurella* vaccine, Fleming's brilliance lay in seeing the breakthrough in an otherwise unplanned situation. Pasteur famously stated that 'Chance favours the prepared mind', which is echoed by Fleming's 'The unprepared mind cannot see the outstretched hand of opportunity'; both of them were saying that only by knowing what went before can you make the lucky breakthroughs. Luck isn't unique to infectious disease research: Velcro, X-rays, microwaves and Post-it notes all came about thanks to a lucky observation.

Luck or serendipity, to make it sound more scientific, contributed to the discovery by Robin Warren and Barry Marshall that infections not stress caused gastric ulcers. Having observed that the guts of patients with ulcers contained bacteria, they tried to culture these bacteria; they were unsuccessful until they left them in the incubator over the long Easter weekend, returning on the Tuesday to find colonies. Marshall then famously proved the link by deliberately infecting himself with *Helicobacter pylori*, giving himself both an ulcer and a Nobel Prize. But I wouldn't necessarily recommend self-experimentation; in an infamous study, Michael Smith at Cornell administered a bee sting to twenty-five different parts of his body, five times a day for thirty-eight days.[2] Unsurprisingly, he found that stings to the shaft of the penis were the most painful. No one except him knows why he did this research, though it did earn him an Ig Nobel award, presented for the most ridiculous science of the year. Other winners include the first scientifically recorded case of homosexual necrophilia in ducks, a study of farting herrings and a demonstration that cracking knuckles does not cause arthritis[3] (the author only cracked the knuckles on one hand every day for fifty years).*

The fungi contaminating Fleming's plates was a *Penicillium*, from which he named the active compound penicillin. Nothing much happened between the initial discovery in 1928 and 1940, when Howard Florey and Ernst Chain (latterly of Imperial), while working at Oxford University, developed a process for extracting penicillin from fungal broth, for which they shared the 1945 Nobel Prize with Fleming. Working in the fraught early days of the Second World War, Florey and Chain rubbed spores of *Penicillium* into their clothes, so if forced to flee they could continue their research. Florey and Chain's method was effective, but not efficient enough to produce a drug in sufficient quantities to treat people, because the strain of *Penicillium* they used, *P. notatum*, didn't produce much penicillin. Mary Hunt (or Mouldy Mary), working for the US Department of Agriculture in Peoria, Illinois, had the romantic job of searching local markets to find better strains. In 1943 she struck mould. She found a strain of *P. chrysogenum* on the skin of a going-off cantaloupe, which produced two hundred times as much penicillin. A bit like selecting cows for greater milk yield, the scientists

* The article in question has one of the best openings of any scientific paper: 'During the author's childhood, various renowned authorities (his mother, several aunts, and, later, his mother-in-law [personal communication]) informed him that cracking his knuckles would lead to arthritis of the fingers.'

selectively bred a more productive yeast by bombarding it with X-rays.[4] Manufacturing scale-up followed and 2.3 million penicillin doses were prepared in time for the D-Day landings in 1944. Soldiers being soldiers, many of these doses were used to treat soldiers who had acquired STIs rather than battlefield injuries. In an unfortunate twist to Churchill benefiting from the German-developed sulfa drugs, Hitler was most likely treated with Allied penicillin following the von Stauffenberg bomb plot.

Following the war, research on novel antibacterial drugs intensified and the twenty-year period from the end of the 1940s to the 1960s is often called the golden age of antibiotics. Half of the antibiotic drugs currently in use were discovered in this time period, including chloramphenicol, tetracycline, cephalosporin and vancomycin. Selman Waksman, a Ukrainian-born microbiologist, spearheaded this work while working at Rutgers in New Jersey. Rutgers is an institute dear to my heart, because it was where I began, and ended, my international rugby career – admittedly only 15 students from a cohort of 67,000 played rugby. Waksman coined the phrase antibiotic and his lab generated many of the antibiotics still in use today.

In 1943, Waksman, or more accurately Albert Schatz who worked in his lab, was testing different compounds for their effect on the growth of *Escherichia coli*. *E. coli* colonises the gut, with variable levels of aggression from mild commensal to deadly pathogen – for example, in 1996 twenty-one people died in Wishaw, Scotland after sharing a church meal made from contaminated meat. *E. coli* also causes nearly 80% of urinary tract infections (UTI), the most common form of bacterial infection. *E. coli* isn't all bad, however: it is the workhorse of molecular biology, frequently manipulated for genetic research – Genentech used *E. coli* to make recombinant insulin.

To try and kill *E. coli*, Schatz and Waksman replicated the accidental approach of Fleming in a more systematic way – they looked for compounds made by one bacterial species that could kill another. As described in Chapter 4, which deals with the microbiome, bacteria are fiercely competitive with one another in a bid to get access to the resources in any environment. Eventually Schatz and Waksman isolated streptomycin from a species of bacteria called *Streptomyces*. Streptomycin could kill a broad spectrum of bacteria, including TB, thus making it the first anti-TB drug; a significant breakthrough.

Waksman got the Nobel Prize for his role in antibiotic discovery. But as with many of these high-profile breakthroughs, disputes over ownership of the discovery arose and Schatz ended up suing Waksman for downplaying his role in the discovery of streptomycin. Another lab worker, Elizabeth Bugie, also missed out; her peers told her that she needn't be included, because one day she would get married and have children and not need to worry about publications. Bugie went on to work at Merck, to evaluate the effect of penicillin on TB. The real winner in the story is the bacteria *Streptomyces griseus*, which now has the honour of being the State Microbe of New Jersey, being recognised as such in 2019. New Jersey is not the only state with a microbe: Oregon has claimed *Saccharomyces cerevisiae* (brewer's yeast), Wisconsin *Lactococcus lactis* (cheese mould) and Illinois *Penicillium chrysogenum* (from Mary's mouldy cantaloupe). Brilliantly, the Senate of Hawaii are embroiled in a feud between *Flavobacterium akiainvivens* (found in the bark of a Hawaiian tree) and *Aliivibrio fischeri* (a luminescent bacteria found in squid) – who says politicians don't deal with the important issues of the day?

HOW ANTIBIOTICS WORK

What I find remarkable is that the first antibiotics were identified with no real understanding of how they worked. But it shouldn't be that surprising. Many drugs are effective without us understanding why, including metformin (for diabetes), ketamine (the horse tranquiliser that can also be used to treat depression) and, remarkably, paracetamol. While we like to pretend that we know everything, a lot of the drugs we use were discovered by blind luck – but whisper it quietly or the government might stop all our funding.

As with vaccines, two schools of thought exist when it comes to drug development: the empirical (test and refine) and the theoretical (design from first principles). The empirical approach can be daunting: testing thousands of compounds is no mean feat – though Waksman's approach of setting a student on it and then claiming the credit for the work has its attractions. The empirical approach benefits from millions of years of evolutionary refinement, especially when natural compounds are used as a starting point. And as described above, lack of knowledge about how something works is not a barrier to making something that does work. The theoretical approach is more attractive at some levels, because it means we can carry on tinkering in our labs but pretend that the work we do has 'translational' value, beyond the esoteric value of science for science's sake.

We do now know a bit about how antibiotics work. There are two broad groups of antibiotics: bactericidal antibiotics (those that actively kill the bacteria) and bacteriostatic antibiotics (those that stall bacterial growth, giving the immune system time to kill them off). In broad principle, antibiotics work by targeting a unique aspect of the bacterial life cycle.

For example, penicillin targets the packaging surrounding bacteria (see figure 11A). Bacteria are surrounded by a cell wall (a toughened capsule that protects them from the environment). In this respect they differ from human cells, which are surrounded by a cell membrane made mostly of lipids (it is much more fragile, like a soap bubble). When the bacteria grows it needs to make more of the cell wall; it does this by adding more building blocks. To make the wall robust, the bacteria performs a chemical reaction that links the blocks together; an enzyme called DD-transpeptidase catalyses this reaction. Penicillin (because it looks a bit like the building blocks) binds to the active slot of the bacterial enzyme and prevents the strengthening reaction from taking place. In the presence of penicillin, the bacteria makes leaky cell walls and when it tries to grow it pops – it's like a hole in a pair of tights, which looks small until you stretch them and then it rips the whole thing apart. The antibiotics that target this process, including penicillin, are called beta-lactams because of the characteristic chemical structure of the drug; a beta-lactam ring is formed of three carbons bound to a nitrogen in a square with an oxygen atom (see figure 11B).

We know the structure of penicillin and beta-lactams thanks to the work of Dorothy Hodgkin. Hodgkin was born in Cairo to English parents and went to Somerville College in Oxford. Her fascination with chemistry began when her mother gave her a book on X-ray crystallography.

X-ray crystallography is how we interpret the shapes of molecules smaller than the visible eye. The first, and not trivial, step is to make a pure crystal of whatever you want to look at. You may remember hanging a small bit of copper sulphate on a thread in a blue solution and a bigger crystal growing on it. Protein crystallography is a bit like that, but with a complex biochemical. Once the crystal has grown big enough you can then fire X-rays at it and interpret the diffraction pattern. It is not dissimilar to how an X-ray of your lungs works – the X-rays are absorbed by the hard bits of bone but not the soft bits of skin, leaving an inverse image (see figures 11C–D). In the case of a crystal the X-rays bounce off the electrons in the constituent atoms in predictable ways, from which one can infer the structure. Hodgkin excelled at it and she demonstrated that penicillin contained a beta-lactam ring. As well as the structure of penicillin, she solved the structure of Vitamin B_{12} (earning her the Nobel Prize) and insulin, a project that took her over thirty-five years.

Hodgkin has one other legacy; Margaret Roberts, the future Iron Lady, studied in her lab. Chemistry apparently prepares the mind for politics – Angela Merkel, like Thatcher, studied chemistry before leading her country. Reflecting the shortage of MPs with any real experience outside of the Oxford debating circuit and the Conservative Central Office, there is a dearth of scientifically trained MPs in the current Parliament, with just Thérèse Coffey having a science PhD. Admittedly there are more medically trained MPs, which is not necessarily surprising.

While bemoaning the current make-up of the UK Parliament is a fun sport, it doesn't necessarily get us any further in thinking about infectious diseases. Hodgkin's

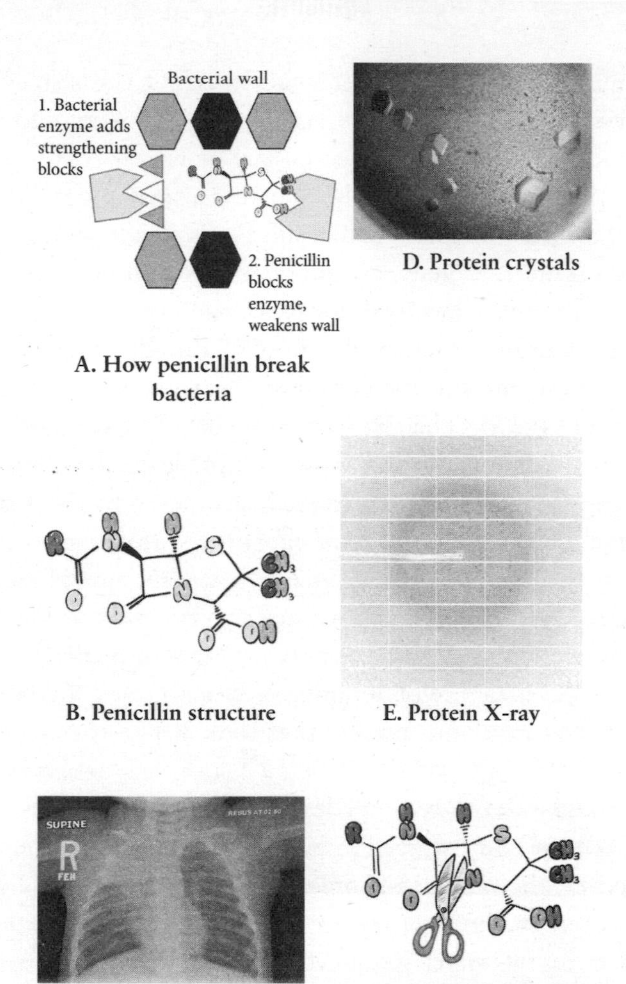

Figure 11 How penicillin works: A. How penicillin kills bacteria. Normally the bacteria strengthen their walls by adding molecules between them – called crosslinking. Penicillin blocks the enzyme that does this, weakening the walls, so when the bacteria grows, it pops. B. Penicillin structure (solved by X-ray). C. Lung X-ray. D. Protein crystals. E. Protein X-ray. F. How bacteria fight back.

finding supported a possible mechanism of action of penicillin. Other antibiotics target other pathways unique to bacteria; for example, chloramphenicol targets the bacterial protein factory, the ribosome, and rifampicin targets the enzyme that transcribes the RNA message from the DNA in the nucleus.

Targeting pathways unique to the bacteria is good news in terms of killing bacteria and not humans, but antibiotics are a blunt tool when it comes to other bacteria. We live in close harmony with our bacteria in the form of the microbiota and all the benefits it provides us. It is not possible to kill only the bacterial pathogen causing you grief, because you often take out vast swathes of other good (or at least neutral) bacteria at the same time. This explains why a course of antibiotics can lead to stomach upsets; it is because it unbalances the communities of bacteria in your guts. In extreme cases, this can cause much more severe disease. *Clostridium difficile* is a common secondary infection that occurs in frail hospitalised patients. The patient will come into hospital for something else, for example a broken hip, and to prevent infection will be given broad-spectrum antibiotics (basically a cocktail to kill everything). The antibiotic cocktail in this case acts a bit like a forest fire; it burns down everything in the guts and, like a forest fire, whatever recovers fastest chokes everything else out. *C. difficile* is normally held in check by the competition from the other bacteria in the guts but after antibiotic treatment it takes advantage of the denuded guts to establish a foothold. The solution to this is simple, but also slightly off-putting – faecal transplant. It is exactly what it sounds like – transferring poop from one person to another. The treatment can be euphemised as FMT (faecal microbiota transplant), to make it more sciencey sounding.

Poop contains all kinds of bacteria. Most of your poo is water (about 75%), but if you were to dry it down, and I am not necessarily suggesting that you do, 30% would be indigestible food, 20% would be fat and 20% salts; the remaining 30% would be bacteria – some dead, some alive. The principle behind FMT is that this 30% can then act as a starter to reseed the guts; it is the medical equivalent of giving someone your sourdough starter. FMT has been so successful in reducing *C. diff* disease that it is now the standard of care (the go-to treatment). But one risk with FMT is that a different pathogen could be transplanted at the same time as the good bacteria. In March 2020, the Food and Drug Administration (FDA) issued a warning about severe *E. coli* infection following FMT. To overcome this, a couple of approaches are being considered: firstly, pre-banking stool from patients prior to planned hospitalisation, for example surgery, so you are repopulated with your own microbiome. Alternatively, a number of companies are developing a more rationally designed cocktail of bacteria which can be given orally as a pill.

THE BACTERIA STRIKE BACK

However, there is a much bigger problem with antibiotics than collateral damage to good bacteria. They are beginning to fail. Pathogenic bacteria are fighting back, gaining resistance to drugs. We are seeing a huge rise in antibiotic resistance, which is not as it might sound: it does not mean resistance of people to antibiotics, but the ability of certain bacteria to survive antibiotic treatment. It is also different to antibiotic intolerance, when people experience an allergic-like reaction to antibiotics. Drug resistance is a serious problem: each year half a million people die around the world because of antimicrobial resistant (AMR) bacteria, which amounts to an eighth of the COVID-19 pandemic, *annually*. Some strains are resistant to only one drug, but more worryingly there are multi- (MDR), extensively (XDR) and pan- (PDR) drug resistant bacteria. The problem isn't necessarily a new one, because Fleming predicted the emergence of antibiotic resistance in 1945. If we return to the time course of antibiotic discovery we can see how quickly resistance emerged (see figure 12), often emerging during the first clinical trials of the new drug. Within ten years of the first trial of penicillin, 50% of *S. aureus* strains were resistant to it.

Resistance occurs in two ways: *de novo*, where a mutant bacteria emerges in a population that can fight off the drug, or through pre-existing mechanisms that bacteria have evolved to protect themselves – which makes sense, as most antibiotics are derived from natural products.

The mechanisms of resistance are as varied as the drugs themselves. Returning to our old favourite, penicillin and its beta-lactam ring, resistance to this drug comes in the form of a bacterial gene called beta-lactamase (see figure 11F). Scientists use the suffix -ase to denote enzymes that can break down other substances; for example beta-lactamase breaks down beta-lactam. Ernst Chain and Edward Abraham identified the first beta-lactamase, penicillinase, in 1940, which is mind-blowing if you think penicillin was not widely in use until 1944. Other mechanisms include pumps that dump the drug out of the bacteria, thicker cell walls that stop it ever getting in and alterations to the bacterial protein that the drugs target.

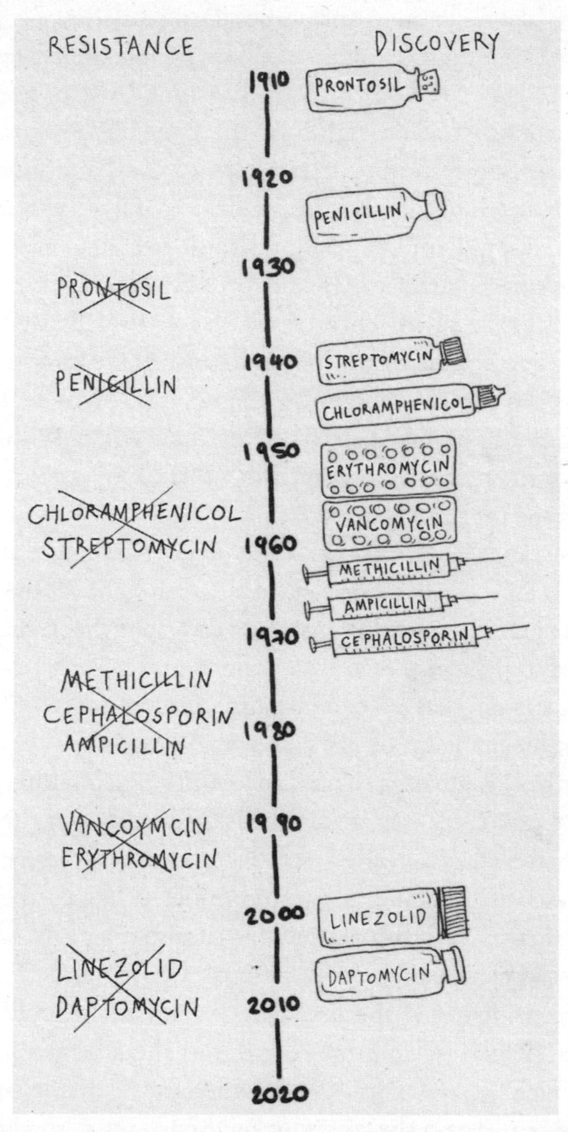

Figure 12 A history of resistance: Bacteria resistant to antibiotics have emerged almost as quickly as the drugs themselves were developed.

The problem is exacerbated because, surprisingly, bacteria can copulate, spreading resistance around. Bacterial sex is not wildly different to animal sex; one bacteria uses a long rod-like structure (called a *pilus*, from the Latin for hair) to poke into another bacteria and pass genes across. Esther and Joshua Lederberg performed much of the work that underpins our understanding about bacterial sex. Esther was the practical experimentalist, developing novel methods for culturing bacteria, and Joshua the theoretician – it was a symbiotic relationship; neither would have thrived without the other. While both Lederbergs received a Pasteur Award for their work, in 1958 Esther had to watch on as her ex-husband Joshua won the Nobel Prize for the work they had done together duc to the inherent bias of the time.[5]

Bacterial genes are mostly transferred on something called a plasmid: a mobile DNA element that bacteria use in addition to the main source of genetic information in their chromosomes. Plasmids are always presented as perfect circles in textbooks and gene maps, but real life being what it is they are actually messier, looking like an elastic band that has been dropped into a drawer. This method of passing information between bacteria has been borrowed by scientists to insert genes into bacteria; for example, the gene for insulin.

Antibiotic resistance mostly comes at a cost to bacteria. The cost may only be marginal; for example, having one extra gene means that the bacteria has to make a bit more DNA each time it replicates, which will make it grow fractionally slower than another bacteria that doesn't make the gene. In the absence of an antibiotic, the sensitive strain, because it grows slightly quicker, will out-compete the resistant strain. But in the presence of the antibiotic the resistant strain lives and the sensitive strain dies.

How have we got into the state where we are frittering away our most valuable medicines? While resistance is natural, its increasing prevalence has roots in human behaviour. The emergence of antibiotic resistance is multi-factorial, with both biological and societal causes, but it all comes back to overuse of antibiotics. Every course of antibiotics selects for resistant strains: it is a classic survival of the fittest.

One major problem is the size of the human population. Even if prescribed and used properly, 7.8 billion people require a lot of doses of antibiotics in any given year. A lot of the drugs given will pass through people unaltered, ending up in sewage treatment works which are now hotspots of antibiotic-resistant bacteria.

But antibiotics are not always prescribed and used properly. Some of this is the fault of us as consumers (and parents/carers). Patients put a great deal of pressure on doctors to prescribe something, anything, even though most infections are self-resolving. Likewise viruses, on which antibiotics have no effect, cause many common infections. If you or your loved ones catch a cold caused by a virus, then a course of antibiotics will do nothing for you. It will potentially have a negative impact, because antibiotics damage your microbiome.

While most doctors do not prescribe drugs unnecessarily, doctors are sadly not the only source of antibiotics and it doesn't take very long to find antibiotics online (on 27 September 2020, Google found me 1,150,000,000 results in 0.59 seconds). Tragically, in low- and middle-income countries where rates of infection with antibiotic-resistant bacteria are higher, access to antibiotics is often less well controlled and the antibiotics sold may be of lower quality, creating a downward spiral.

Once prescribed, there is still scope for misuse; through failure to complete courses of antibiotics, hoarding of left-over antibiotics and inappropriate disposal. Not using up the whole prescribed course of antibiotics is bad – you will fail to kill all of the sensitive bacteria and enable the resistant ones. The public health message is simple: finish your course of antibiotics, even if you feel better. Not doing this puts everyone at risk in the long run. The impact of incomplete courses of antibiotics is worse for infections that require lengthy treatment courses, for example TB. There is an ongoing crisis in TB management, with many of the first-line drugs being no longer effective. Many of the resistant strains emerged back in the USSR. The gulags, on top of everything else, were rife with TB. While treatment was available it was limited, so patients received shorter, less effective, courses of drugs. Following the breakdown of the Soviet Union in the early 1990s, two things happened: the health system collapsed and prisoners from the gulags dispersed, spreading the drug resistant bugs far and wide.

If there are any old antibiotics lurking in the back of your drugs cupboard, don't hoard them in case you are feeling a bit ill in the future – they won't work, and they will encourage resistance. Dispose of them properly and look out for antibiotics amnesties when you can hand them into pharmacists, who will dispose of them safely.

Preachy public health messages:

1. Antibiotics won't kill viruses.
2. Only take antibiotics if they have been prescribed.
3. Finish the course of antibiotics.
4. Dispose of them properly (return to pharmacist).

Overuse is not just a problem in people. In 2010, 63,000 tonnes of antibiotics (sixty million kilograms) were given to animals. That's not just to Fluffy the cat with his hurty paw, but in industrial quantities to intensively produced livestock.[6] Many dairy cows are given prophylactic antibiotics to prevent mastitis, getting antibiotics even though they are not sick; and 10% of beef cattle get antibiotics in their feed as a way to accelerate growth. God only knows what goes into chickens and pigs, but it's safe to say it isn't all happy organic corn and sunshine. Cheap meat will kill us all in the end, which is a problem, because bacon is so tasty.

POST-ANTIBIOTIC APOCALYPSE

A post-antibiotic world will be a pretty grim place. Antibiotics have had an incredible impact on human health – we will really miss them when they are gone. There is the obvious impact on treating infections; estimates suggest that penicillin alone has saved 80 to 200 million lives. Bacteria are everywhere; infection is common. Coolidge junior's stinky sneaker is just one example of bacteria getting under your skin. Bacterial infection will be a problem for everyone and doubly so for immunocompromised people, where every bacterial infection is potentially fatal.

An illustration of the impact of antibiotic resistance (AMR) is the rise, fall and rise of infection after blast injuries. Infection has always been a major contributor to death during wartime. Gangrene has a place in the First World War's pantheon of horrors; it was sustained when bacteria colonised wounds following artillery bombardment, with a lethal cocktail of metal, cloth and dirt being introduced into bodies at high velocity. This would lead to a creeping infection for which the only treatment at the time was progressive amputation. The advent of penicillin meant that the rate of gangrene was 0.15% in the Second World War, compared to 14% in the First World War. Infections after blast injury still occur – firing sand, mud and metal into the body at force is still an excellent way to introduce bacteria at the same time. Unfortunately, bacteria introduced this way are increasingly resistant to penicillin. Strains of bacteria commonly associated with blast injuries – *Staphylococcus aureus* and *Acinetobacter baumannii* – now exist that cannot be killed with any of the available antibiotics, raising the spectre of gangrene. One approach to get around this is to make vaccines against them and Sophie Higham in my lab is working on this as part of her PhD.

Antibiotics have contributed to an unparalleled age of medical advances. Surgery in general is a risky activity – opening up the body to the outside world and introducing foreign objects, such as hips, is party time for bacteria. Intubation and catheterisation, both of which are critical for intensive care support, are also great vehicles to bring the bugs back home. The process of transplantation requires the host immune system to be shut down, which dramatically increases susceptibility to infections and is only manageable thanks to antibiotics. The dramatic increase in the lifespan of cystic fibrosis patients is also down to antibiotics, as is the survival of preterm infants in neonatal units and trauma victims in intensive care. None of these things would be possible without antibiotics.

What can be done? There are two broad strategies: get more new antibiotics and preserve the existing ones. In this modern, post-genomics revolution age, finding new ones feels like it should be an easier option. After all, in the sixties they were pumping a new one out nearly every other year. One problem is finding them in the first place: new antibiotics don't just grow on trees – they are found in the soil and on melons.* Finding them takes the investment of time and money and this is drying up. In the 1990s eighteen drug companies had antibiotic research and development programmes, but by 2018 only Merck, Pfizer and GSK remained in the game. The reluctance to initiate costly research and development programmes stems from the poor return on investment in this area. The paradox is that we need new antibiotics that we then don't use – save for the most drug resistant infections – which makes them much harder to sell and recover any investment put into them in the first place. Even if scientists develop a nuclear option that can kill all bacteria it wouldn't necessarily be a good idea to use it, because of the speed at which bacteria develop resistance. One way round this economic problem is to get governments to pay for novel antibiotic programmes. This approach of conserving power for a rainy day is, like preventing climate change or investing in good public health, quite easy for governments to kick down the track until it's too late. Governments find it hard enough to spend money on stopping the things that are killing us now – like guns, cigarettes and doughnuts.

Which is not to say there have been no new antibiotics. In 2015, Kim Lewis and his team working in Boston, USA published the first in a new class of antibiotics. Lewis returned to the approach that Waksman, Schatz and Bugie

* Which as we all know grow on the ground, not trees.

used – looking in the soil.* But rather than growing the soil bacteria on artificial media, which can restrict the growth of many microbes, they used a device called the iCHIP, which grows the bacteria in the soil whence they came. They load up the iCHIP with a slurry of soil and then bury it back in the ground to see what will grow. Then they test the internecine tendencies of the recovered bacteria – can they kill other bacteria? Out of the ten thousand strains they found one killer; they christened it *Eleftheria terrae* (meaning free from the earth) and the compound it produced Teixobactin.[7]

But sadly, the path from discovery to clinic is not smooth. Clinical efficacy trials to show that drugs work are not the easiest or cheapest things in the first place and they are even more difficult for novel antibiotics. The numbers of patients are often limited – hospitals hate AMR outbreaks and do their best to prevent them from happening in the first place. The new drugs also need to demonstrate non-inferiority to existing treatments: unlike the physicians in *House*, most doctors prefer not to use experimental medicines in life-threatening situations. All is not completely lost, however; there are some new drugs that make it through. In 2019 the FDA licensed Fetroja (Shionogi), Lasvic (Kyorin), Recarbrio (Merck) and Xenleta (Nabriva). But they target very specific bacteria and have only been licensed for use as emergency drugs of last resort.

We therefore need to preserve the current generation of drugs. Data from Professor Jodi Lindsay (and others), working at St George's Hospital in London, shows that if you restrict antibiotic usage you reduce the spread of antibiotic resistance. Novel approaches to control the use of

* Who should really have formed a band.

antibiotics are going to be key. Professor Alison Holmes, a clinical researcher working with a cross-disciplinary team at Imperial College, has been exploring the use of microneedle patches that can monitor the level of antibiotic in the blood, ensuring doses are high enough to control the infection but are not wasteful. Other strategies can have an impact: vaccination, by preventing the disease in the first place, means that antibiotics are never required. In one study, vaccinating against *S. pneumoniae* reduced the cases of antibiotic-resistant bacteria by 67%.

Further research into strategies to ensure antibiotics can still be used is vital. Otherwise, we risk returning to a time where poor sock hygiene is a potential death sentence. One thing that is certainly true is that antibiotics don't work for viral infections. For those we need an entirely different class of antiviral drugs and the next chapter will explore their discovery and how they work, focusing on the HIV crisis and how drugs saved the day.

CHAPTER 11

Antiviral

Timeline: Late October 2020. Carne, Cornwall. UK lockdown two imminent. Global COVID-19 cases 44,888,869; deaths 1,178,475.

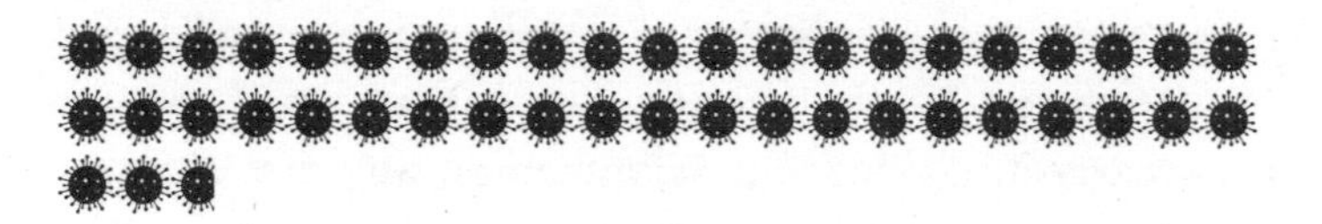

'What will I be doing in 20 years' time? I'll be dead, darling! Are you crazy?'

Freddie Mercury, 1984

On 24 November 1991 Freddie Mercury died of AIDS, the disease caused by HIV. At the time the virus was untreatable and HIV infection was often described as a 'death sentence'. However, in the thirty years since that day antiviral drugs have completely changed the face of HIV. Where anti-retroviral treatments are available and used, it is now a chronic condition that can be managed into old age. We have seen some astonishing successes in the treatment of viral infections – not just for HIV, but also for HCV, another previously untreatable virus.

The history of antiviral drugs is somewhat shorter than the history of antibiotics, for the same reason that antiviral vaccines have been around for a shorter amount of time – viruses were discovered later than bacteria. The first licensed antiviral drug was IDU (or 5-iodo-2′-deoxyuridine to its friends). IDU is a nucleotide analogue. It looks enough like one of the building blocks of DNA (thymidine) that it can get incorporated into a DNA molecule. The iodine part of IDU (the 5-iodo) stops more DNA building blocks from being added to the chain, similar to Fred Sanger's chain-termination sequencing method. Both host and viral DNA incorporate IDU, but while the host cell can repair broken DNA the virus cannot and so the drug specifically kills viruses. The same general principle for how antibiotic drugs prevent bacterial infections applies to antiviral drugs – they target something unique to the virus that the host cell doesn't use (see figure 13). Identifying unique features of the viral life cycle is critical to the development of antiviral drugs.

The demonstration that IDU, a nucleotide analogue, could treat viral infections sparked multiple programmes of research to develop antiviral drugs; one of which, still in use today, is **acyclovir**. Acyclovir outperforms earlier drugs like IDU because of its reduced toxicity, which is a consequence of it being a prodrug, like the antibiotic Prontosil. Prodrugs only become activated in the targeted cell, rather than being active from the moment you take them; other examples include aspirin, codeine and psilocybin (which puts the fun in fungi). Because it needs to be switched on by a viral enzyme called thymidine kinase, acyclovir is only functional in infected cells, which reduces the toxicity and increases the specificity. An oral version of acyclovir, valacyclovir, is commonly used for cold sores – the unsightly spots that appear around the mouth following herpes simplex virus (HSV) infection. HSV comes in two flavours: HSV-1, commonly associated with cold sores,

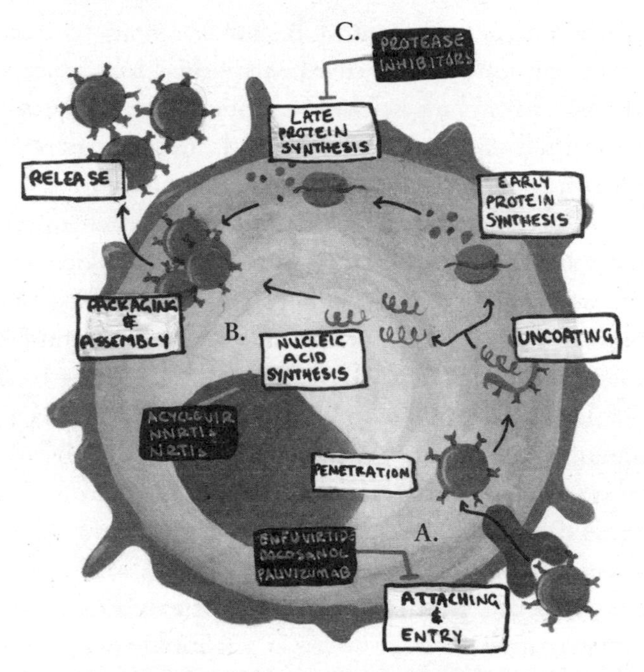

Figure 13 The viral life cycle and how antivirals work: Antiviral drugs target different stages of the viral 'life' cycle. A. Starting at entry, some drugs stop the virus ever getting into the cell. B. Other drugs mimic the building blocks of nucleic acid and stop the virus making more copies of its genes – these are called NRTIs (nucleoside reverse transcriptase inhibitors) in the context of HIV. C. A third class of drugs stops the virus making the proteins it needs to build itself.

and HSV-2, commonly associated with genital herpes (though both viruses can cause either condition). Nearly 67% of the world's population under fifty has HSV-1, making it one of the most common sexually transmitted infections.

Another viral infection that can cause disfigurement is human papillomavirus (HPV), which causes warts and verrucas. HPV warts are commonly treated with salicylic acid (the main component of Bazooka!), which is loosely antiviral but much less focused than other drugs. Salicylic acid destroys the infected tissue, hopefully taking out the virus at the same time. Treating viral infections by killing the tissue can also be done with something as crude as duct tape; other options include freezing them off (cryotherapy) and cutting them off (surgery). The difference between salicylic acid and acyclovir is the specificity for the virus. Salicylic acid acts like a blunderbuss and acyclovir more like a laser.

Acyclovir laid the groundwork for a new generation of rationally designed drugs, pioneered by Trudy Elion and Howard Schaeffer at Burroughs Wellcome in the US. Trudy Elion is the grande dame of antiviral drugs. She graduated in chemistry in 1937 but as a female scientist was unable to find a research job, so she had to save up to do a master's degree part-time while working as a schoolteacher. Being a woman also prevented her from getting PhD funding, but luckily for the world she got a job working at Wellcome in 1944, exploring nucleotide analogues. At Wellcome she worked with George Hitchings, with whom she shared the Nobel Prize in 1988. They used a targeted approach, designing drugs based on the understanding of how viruses replicate. In this regard the drugs differed from the antibiotics, which were mostly discovered and not made – such as Waksman and his mud and Fleming and his mould. Elion and Hitchings deliberately synthesised chemicals that looked like natural compounds. Over her career, Elion worked on a staggering range of drugs, including allopurinol for gout and trimethoprim for bacterial infections. One of the other viruses she worked on was HIV.

HIV (HUMAN IMMUNODEFICIENCY VIRUS)

Antiviral drugs didn't really come into their own until the HIV/AIDS epidemic. In 1981, the *New York Native* newspaper reported the emergence of an 'exotic new disease' – AIDS (acquired immunodeficiency syndrome). One of the hallmark features of AIDS is infection with pathogens that a healthy immune system would be able to fight off. It was initially observed in gay men and intravenous drug users, who were turning up in hospital with a rare form of pneumonia caused by a fungi called *Pneumocystis jirovecii* (previously called *P. carinii*). Pneumocystis pneumonia had previously only been seen in people with failing immune systems. Simultaneously, more gay men were admitted to hospital with a rare form of viral skin cancer called Kaposi's sarcoma. Both Kaposi's sarcoma and *P. jirovecii* are aggressive and horrible and early AIDS patients had disfiguring welts all over their faces. The US CDC set up a task force to identify the causative agent and after various attempts to name the disease, including GRID (gay-related immune deficiency) and 4H (because it infected heroin users, homosexuals, haemophiliacs and Haitians), they finally settled on AIDS.

However, the causative agent for AIDS remained unknown. The election of Ronald Reagan in 1980 on a wave of conservative Christianity hindered progress, with many of his supporters viewing homosexuality as a sin and AIDS as God's punishment. The CDC, as a government agency, had to recruit help from outside its ranks. It approached the virologist Robert Gallo, who, while working with Doris Morgan, had discovered a protein called interleukin-2. Interleukin-2 could keep T cells alive in culture, which meant that like HeLa cells they could be used to culture viruses – especially those viruses that grew only in T cells.

This discovery of IL-2 and the ability to culture T cells set the scene for Gallo's lab to discover the first human retrovirus – HTLV-1. Retroviruses are a family of viruses with a unique life cycle. Unlike our cells, which use DNA, the genetic material of retroviruses is encoded by RNA. RNA genomes are not uncommon in viruses – the unique feature of the retroviruses is that they can convert their RNA into DNA. Having turned their genes into DNA, they then sneak them into the host cell's DNA. This process, called reverse transcription, has led to nearly 8% of our genomes being viral in origin. The enzyme that performs this reaction, reverse transcriptase, is the one used in the viral PCR reactions described in Chapter 7. The retrovirus Gallo's lab discovered, HTLV-1, can cause leukaemia by inserting its DNA indiscriminately into the T cell's genome and in so doing disrupting the processes that control the replication of the cell, turning it cancerous.

The clinical evidence from AIDS – in particular the frequency of rare infections – suggested that HIV could infect CD4 T cells, because they are the master coordinators of the immune response. In 1984, Gallo and his team published in the journal *Science* that they had identified the causative agent of AIDS, a virus that infected human CD4 T cells, which at the time they called HTLV-3.[1] This was the first study to demonstrate a causal link between the virus – now known as human immunodeficiency virus (HIV) – and AIDS. However, Gallo did not isolate the HIV virus first. A year earlier, in 1983, Françoise Barré-Sinoussi, working in the lab of Luc Montagnier at the Institut Pasteur in Paris, actually discovered HIV – but called it lymphadenopathy associated virus and didn't quite have the evidence to show the causative link with AIDS.[2]

These papers triggered a heated debate between the American and French labs over who had first demonstrated HIV as the causative agent of AIDS. This debate had commercial as well as academic consequences, discovery being linked to ownership of the virus detection test. While the French and American governments eventually settled the patent dispute, the Nobel Prize committee came down on the side of the French team, awarding the Nobel Prize to Montagnier and Barré-Sinoussi but not Gallo. Whether this was fair or not is above me, but both labs clearly contributed to the discovery: Montagnier's lab used tools developed by Gallo to grow enough CD4 cells to culture HIV and Gallo's lab used samples from Montagnier.

So, where did HIV come from? The most likely answer is monkeys. A closely related virus, SIV (simian immunodeficiency virus), can be found in primates. A version of this virus probably passed from monkeys to man. The most likely hypothesis is that someone hunting primates for meat was infected by contaminated monkey blood. While most people are immune to SIV, underpinning immunodeficiencies in the exposed hunters could have enabled the virus to take hold and adapt to the human immune system. There are four main strains of HIV (M, N, O and P), which suggests this jump from monkey to man occurred multiple times. The earliest confirmed case of HIV was detected in a blood sample collected in 1959 and retrospectively analysed. By building family trees of the virus and projecting backwards it can be estimated that HIV most probably emerged around Kinshasa in the Democratic Republic of the Congo in the 1920s.[3] This retrospective projection can be made because the rate of viral mutation is reasonably constant and therefore the number of mutations from a single common ancestor can be used to calculate back to when the first case occurred. In *The Origins of AIDS*, Jacques Pépin describes how population growth, a thriving sex trade and railways accelerated the spread outwards from Kinshasa.[4] The virus then hitched a ride across the Atlantic in Haitian teachers who had been working in Kinshasa to replace the void left by the collapse of the Belgian colonial infrastructure in the 1960s and 1970s. It remained undetected in Haiti due to the very poor healthcare system, but probably spread from Haiti (as a destination for gay American sex tourists), priming the outbreak in the USA in the early 1980s.

The discovery of the virus that caused the AIDS epidemic allowed the development of tests to diagnose it and drugs to treat it. At this point, Dr Anthony Fauci enters the scene. Remarkably Dr Fauci, born in 1940, has been head of the US National Institute of Allergy and Infectious Diseases since 1984. He led the response to both the HIV and Ebola outbreaks and estimates he has appeared in front of Congress more than 250 times. In 2020 he was still head and you will most likely know him as the voice of reason in a White House gone mad: at one point he was caught facepalming during Trump's tirades on the coronavirus. In the 1980s, Fauci redirected the NIH research programme to become the world's largest funder of HIV research. Research was critical in the fight against the virus and this is where Trudy Elion reappears. Her research at Burroughs Wellcome on nucleotide analogues paved the way for the first anti-HIV drug AZT (also called zidovudine). AZT had originally been synthesised back in 1964 as an anti-cancer drug, but it was shelved because it didn't work. The only thing AZT could treat was a retroviral infection in mice called Friend's virus, but at the time (1974) there had been no reported disease in humans associated with retroviruses, so AZT was shelved again. In 1985, AZT re-emerged as one of a panel of drugs repurposed for testing against HIV. Repurposing is not uncommon – for example, the potential anti-coronavirus drug remdesivir was originally developed as an anti-Ebola virus treatment. AZT was shown to have anti-HIV activity in cells and then in people in rapid succession.[5] Based on this data, the FDA approved the drug in 1987 – twenty-five months after the first demonstration of action in the lab, which is incredibly fast.

Burroughs Wellcome patented AZT in 1985, which brings us, not for the first time, to the issue of pharmaceutical companies and the 'market' for drugs against infection. For now, let's add an 'it's complicated' as a placeholder and move on, coming back to it in the next chapter.

Sadly, AZT on its own was not the silver bullet everyone had hoped for. The problem is that HIV mutates, rendering AZT ineffective. HIV stores its genetic material on RNA and has a higher rate of mutation because when it replicates it doesn't check the copies, so errors crop up. HIV is a master of mutation: as described in Chapter 5, mutation is a built-in feature of HIV that it (like other viruses) uses to escape the immune system. Once HIV has infected a CD4 T cell, the infected cell alerts the immune system that it is infected by displaying bits of the virus on the MHC. This flags the cell for destruction by CD8 T cells. To escape the immune system the virus changes shape and the CD8 T cells are no longer able to see it. This is classic natural selection: a virus that escapes the immune system can make copies of itself for future generations; the ones that are killed do not, so over time the virus evolves.

The newly evolved viruses are still visible to the immune system, as we have lots of different CD8 T cells that can see lots of different things. But over time the virus can completely escape the immune system, because there are more different conformations of viral proteins than there are CD8 cells that can recognise them. When the virus finally finds a conformation invisible to the immune system, it has free rein to infect and kill CD4 T cells at will. This viral destruction of CD4 cells tips HIV-infected individuals into AIDS, making them susceptible to the hallmark secondary infections such as pneumocystis pneumonia. The rate at which this process of viral escape and T cell destruction occurs depends upon the HLA genes of the infected person. Some HLA types are associated with good viral control and others with bad. It is possible, though rare, to be infected with HIV and never develop AIDS. These people are described as long-term non-progressors, though they make up less than 5% of the total number of people living with HIV.

While the process of viral mutation described above evolved as an adaptation to escape the immune system, it also enables viruses to escape drugs. Drugs kill off the susceptible viruses but leave the resistant viruses, which can then make copies of themselves. We see this with AZT when used as a single drug monotherapy. The drug AZT targets reverse transcriptase, the enzyme HIV uses to turn its RNA genome into DNA. Mutations occur in the viral reverse transcriptase, which improves its discrimination between the AZT drug molecule and thymidine (the nucleotide building block that should be inserted into the DNA), rendering the drug ineffective. With a single drug, mutations can occur quickly. Fortunately, there is a solution to the problem of drug resistance, which is the use of a drug cocktail, each drug targeting different parts of the viral life cycle; including entry into the cell, replication of the genetic material, insertion of the genetic material into the host genome and cutting the viral proteins into active forms. There are more than twenty-five licensed drugs for HIV and combining them delays HIV's ability to mutate and escape. When used together, these drug cocktails or highly active antiretroviral drugs (HAART) are a game changer.

There has been a remarkable turnaround from the early 1980s. Professor Sarah Fidler, a clinician and researcher, has worked on HIV infection at St Mary's Hospital in London since the early 1990s. Training as a junior doctor in the HIV clinic set up by Willie Harris and working with Professor Jon Weber, she encountered the peak of the disease in the UK. She described to me how the clinic had about forty beds filled with young, mostly gay, men; all of them dying. Weber recalled that there was a patient dying each day at St Mary's. AZT sadly made no impact on this, extending life only by three to six months. Fidler remembered the terrible stigma of the disease, wrapped up in the prevailing anti-homosexual attitude of the time. But she also recollected looking after patients on one of the wards when Princess Diana came in to chat to them, famously shaking the hand of a man with HIV, and how this helped remove some of that stigma. In 1992 Fidler left the wards for six years to undertake research for a PhD. It was in this six-year period that HIV treatment underwent a seismic shift. This metamorphosis built upon clinical trials performed by Weber and his team, which demonstrated that using two drugs together could stop patients with HIV developing AIDS. Fidler told me about returning to the wards to see across the way a handsome, healthy, muscular man, who looked strangely familiar: she realised that the last time she had seen him was when he was emaciated and dying in the clinic. His was just one of many lives saved by HIV drugs. Thirty-three million people have died of HIV/AIDS since the start of the epidemic, but thirty-eight million people with HIV are still alive. HAART saved approximately 1.2 million lives in 2016 alone.

HAART doesn't just save lives: it can prevent infection in the first place. One of the many tragedies of HIV is that it can pass from mother to child. This can happen in the womb, during delivery or while breastfeeding. Babies born with HIV only have a life expectancy of ten years if left untreated. However, when mothers are tested and treated during pregnancy, the use of HAART prevents transmission to the unborn child, breaking the chain.

Using drugs to reduce the level of the virus in people with HIV can also reduce the risk of onwards transmission. Lowering the viral load in infected patients means that they are in turn less infectious. Initially, HIV drugs were reserved for individuals whose viral load had skyrocketed and whose CD4 T cell count had crashed. But thanks to research by Professor Fidler and others as part of the PopART study, the drugs are now used much earlier during infection, preserving the immune system and preventing new cases.[6]

Regardless of their efficacy, taking a cocktail of drugs every day for life is far from ideal. The next big thing for HIV research (apart from a vaccine) is a cure. An HIV cure has actually happened twice. This can be looked at in two ways: pessimistic or optimistic. Pessimistic: it has ONLY happened twice, so given that 76 million people have been infected with HIV, two people is a pitiful rate (0.000003%). Optimistic: it HAS happened TWICE; you have to start somewhere and that it can be done and can be repeated is cause for hope. The first person to be cured of HIV was Timothy Ray Brown, also known as the Berlin patient. Brown was diagnosed with HIV in 1995 and was prescribed antiviral drugs, the same as for any other person living with HIV.[7] However, in 2006 Brown reported feeling exhausted the whole time and was diagnosed with acute myeloid leukaemia. Leukaemia is a blood cancer caused by the white cells replicating uncontrollably: myeloid leukaemia is specifically associated with a cell called a myeloblast, the precursor cell for macrophages and neutrophils. Back in Berlin, Dr Gero Hütter initially treated Brown for his leukaemia using chemotherapy. This was unsuccessful and Brown needed a stem cell transplant, where the opportunity for an HIV cure arose.

Stem cell transplant is where the cancerous white blood cells are replaced by someone else's; the new white blood cells then kill the old ones, removing the cancer. The success or not of a stem cell transplant relies on matching the donor and the recipient. If they are mismatched the new cells transferred in from the donor start attacking all the organs of the body in a condition called graft vs host disease. Matching the donor and recipient is very difficult, because there needs to be a genetic match in HLA types, the genes which determine how your T cells distinguish self from non-self. The problem is that unlike simple characteristics such as, for example, eye colour or tongue rolling, which are determined by a single gene, HLA type is determined by six different genes.* Genetic codominance further complicates

* There are three loci for MHC-I – HLA-A, HLA-B and HLA-C – and three loci for MHC-II – HLA-DP, HLA-DQ and HLA-DR.

the situation. For most of our genes we inherit two copies, one each from our mother and father, but normally one of those copies overrules the other; for example, brown eye colour is dominant over blue eye colour, as is tongue rolling. But with the HLA genes, our cells make both the maternal and paternal proteins. This means we make up to twelve different proteins, with a huge number of variants of these genes to choose from. Most stem cell transfers come from siblings, as they are the most likely to inherit similar genes; but there are also donor banks.

Timothy Brown was in the relatively fortunate position of having a number of possible matches in the donor bank. This led Dr Hütter to test out an idea. HIV infects CD4 T cells by binding to two proteins on the cell surface, CD4 and another protein called CCR5. CCR5 belongs to a family of proteins called chemokine receptors, which determine where in the body the immune cells go to. CCR5 is somewhat redundant – other proteins have a similar function and there are people who express a non-functional variant of the protein called CCR5 Delta 32 (CCR5Δ32) who are perfectly healthy. This CCR5Δ32 variant lacks 32 base pairs, which makes it non-functional. Observations in the 1990s suggested that cells with the delta32 variant could resist HIV infection. This observation, that CCR5Δ32 was protective, led Dr Hütter to his breakthrough. He identified a stem cell donor who not only matched Timothy Brown but was also CCR5Δ32. Brown stopped taking his antiviral drugs on the day of his bone marrow transplant and remained HIV negative. When Brown passed away in 2020, thirteen years later, he was still HIV free. In 2019, a second 'London patient' (Adam Castillejo) was revealed as being cured of HIV and at the time of writing he remains HIV free.[8] This approach is not going to be a panacea, as it is very aggressive and quite dangerous for the recipient; Brown required two cell transfers and after the second one experienced severe adverse effects, nearly going blind. Because of these risks, it is unethical to perform a bone marrow transplant in the absence of a disease necessitating it, for example leukaemia.

The promise of these two patients has led to a darker use of science for the prevention of infection. In 2018 Dr He Jiankui, based at the Southern University of Science and Technology in China, gene modified human babies to try and make them HIV resistant. His – weak as piss – rationale was that the parents were HIV positive so it would prevent mother-to-child transmission. Which is BS because drugs are completely effective in blocking the transmission of the virus to the baby. He used the CRISPR/Cas9 technology to target the CCR5 in the babies. The off-target effects of CRISPR/Cas are not fully known, making it entirely possible that in cutting the CCR5 gene out he has altered other genes in the babies. These concerns didn't stop him and in November 2018 he announced the birth of twins Lulu and Nana. Things proceeded rapidly from there – he was suspended from research on 28 November and jailed on 30 December.

Rogue scientists aside, the progress in HIV treatment has been remarkable over the last forty years. But HIV is not the only good news story in viral drug treatment. In 2011, a cure for hepatitis C virus (HCV) was licensed, which has dramatically changed the course of this disease. One of my reasons for writing this book was this quiet miracle, which I suspect, like HIV, has passed under the radar because people infected are often on the fringes of society – IV drug users, prisoners and sex workers.

HEPATITIS C VIRUS (HCV)

Hepatitis is inflammation of the liver, which is characterised by fatigue, weight loss and jaundice. It may ultimately lead to liver failure and death. Hepatitis can have non-infectious causes, most commonly alcoholism. An infectious cause of hepatitis was first identified in the 1940s. Viruses are the predominant cause of infectious hepatitis. There are three types of viral hepatitis, conveniently called A, B and C, though confusingly the first virus to be identified was hepatitis B. It had been seen that viral hepatitis could be transmitted during blood transfusion, so identifying the cause of the infection would help with blood screening. Sadly, other viral infections can be passed on with donor blood and haemophiliacs had high levels of HIV in the 1980s because of their need for more frequent transfusions. Returning to HepB, Baruch Blumberg identified surface antigen in 1963, leading to both the diagnostic test and the vaccine, earning him the 1976 Nobel Prize.

Despite screening the blood banks for this agent there were still cases of hepatitis in patients who received blood transfusions. This led Harvey Alter to hunt for a cause. By infecting chimpanzees with contaminated blood, Alter demonstrated the presence of another hepatitis-causing agent in the blood banks that was neither HepB nor HepA (which had been discovered in 1977). Michael Houghton investigated this contaminated blood at the biotech Chiron, the same company that had developed the HepB vaccine. Houghton cloned random fragments of DNA from infected chimpanzees and screened these using blood from infected patients, to identify which bits of DNA encoded viral proteins. This answered some but not all of Koch's postulates. To truly confirm that the putative hepatitis C virus (HCV) caused the disease, it needed to be shown that infection could be caused by this virus and nothing else. To do this Charlie Rice, working at Washington University, generated a clone of HCV and demonstrated that this could indeed infect chimpanzees. For their work, Alter, Houghton and Rice received the Nobel Prize for Medicine in 2020. As with other Nobel Prizes, it was awarded to the leaders of the labs involved but represented the efforts of a much larger team, which Houghton publicly acknowledged: 'All of these brilliant people deserve recognition.'[9]

The identification of the virus that caused hepatitis C disease led to the development of therapies. The first therapy attempted was interferon. As described earlier, interferon acts as a natural antiviral. Interferons trigger a pathway of proteins that prevent viral replication. Based on this principle, it was used as a treatment for HCV. Unfortunately, viruses are sneaky. Most viruses evolve mechanisms to escape from interferon. HCV encodes a protein that shuts down the interferon-signalling pathway. It's a bit like cutting the wires from a switch: you can turn it on, but nothing will happen. Given interferon has some pretty unpleasant side effects – you feel like you have a cold all the time – and its relatively low efficacy (effective only in about 5% of patients), it wasn't the breakthrough that had been hoped for.

In 2011 came the licensure of the first direct-acting antiviral drugs for use in HCV treatment. Learning from HIV, the drugs are administered as a cocktail and target three viral proteins, NS3/4A, NS5A and NS5B, once again demonstrating that when it comes to gene names, microbiologists and virologists suck. The drugs stop the virus from making new copies of itself. NS3/4A forms a protease, which is an enzyme that chops the viral proteins into active parts and NS5A and NS5B are involved in the replication of the viral genetic material. The roll-out of HCV therapy has had a dramatic impact on the number of cases, which fell from 180,000 in 2015 to 140,000 in 2018, triggering a major campaign to eradicate HCV within the UK.

TIMING MATTERS: THE USE OF ANTIVIRALS FOR ACUTE VIRAL INFECTIONS

One of the many unforgettable moments of 2020 was when Donald Trump caught COVID-19. It goes without saying that I am not a fan. His criminal mismanagement of the crisis led to a great excess of deaths in the USA; especially as it is the richest country in the world. Trump's failings were innumerate, but included downplaying the threat, blaming it on China, putting the stock market first, undermining the public health messaging, holding rallies in infected states, holding press conferences when his campaign secretary was infected, just making stuff up (injecting bleach and all that guff about hydroxychloroquine), Twitter diarrhoea, not listening to Tony Fauci (yes the same one from the HIV crisis) and generally being a bright orange douchebag. That said, the video of him emerging triumphant over the virus from Marine One was pretty cool in a Leni Riefenstahl way.

About the only 'good' he did was to provide a control group to compare against, which was great for the rest of the world because it accelerated vaccine trials (Operation Warp Spread) but terrible for the citizens he was sworn to protect.

Back to his infection. On 1 October 2020 he was reported positive for SARS-CoV-2, the causative agent. The next day he was flown by helicopter to Walter Reed Army Medical Center in Washington, where he was reported to have been given oxygen and a cocktail of drugs: remdesivir (unlicensed [at the time] antiviral) dexamethasone (a steroid), Regeneron (unlicensed [at the time] antibodies that target the virus), famotidine (a stomach ulcer drug), zinc (supplement, possibly antiviral), vitamin D (supplement, possibly antiviral), melatonin (may help sleep) and aspirin (anti-inflammatory and painkiller). Two of these were particularly interesting – Regeneron, which I will discuss in Chapter 15, and remdesivir.

Remdesivir is an antiviral drug made by Gilead,* who also manufactured the first HCV drug, Sofosbuvir, and many of the anti-HIV drugs. The COVID-19 pandemic in 2020 triggered a wave of antiviral drug studies. As with HIV a lot of this involved repurposing old drugs which had proven efficacy against other viruses. These had variable success. But even with a sizeable library of compounds, successfully targeting a new virus is tricky work. Remdesivir was initially developed to prevent HCV but was then initially repurposed to prevent Ebola virus; and it is not particularly effective against either. Remdesivir belongs to the same family of drugs as acyclovir; it is a prodrug metabolised to make a nucleotide analogue that prevents the virus from making copies of its genes. To test whether remdesivir could reduce disease during COVID-19, a large phase III randomised trial was performed on adults with lung infections.[10] The study reported that patients treated with the drug had a significantly shorter course of disease,

* Gilead is named after the Balm of Gilead, a medicinal perfume from the Bible; nothing to do with Margaret Atwood.

suggesting it was effective. However, another report combining multiple studies suggested that remdesivir had minimal effect,[11] so its efficacy remains unclear.

While they can kill the virus under laboratory conditions, the big problem with antiviral drugs against acute infections is the timing of administration: they are probably most effective when given early during the disease. For COVID-19 and other respiratory viruses, the immune system causes much of the damage, with the virus mainly acting as the spark. Since the virus primarily acts as the trigger for the disease, targeting it once the disease has started is a classic case of shutting the stable door behind the horse.

The difficulty of using drugs against an acute infection has also been seen with influenza. In 2009, the world's population faced the first pandemic outbreak of the twenty-first century: swine flu. In response to this many countries stockpiled antiviral drugs, in particular a drug called oseltamivir, or Tamiflu. For example, the UK government purchased around £424 million pounds' worth of the drug. There was a lot of concern about this purchase; it certainly didn't help that £74 million pounds' worth had to be written off because of poor record-keeping. A bigger problem arises when the data from multiple clinical trials is combined; there is little evidence that Tamiflu has any impact on the outcome of flu infection.[12] The analysis was performed by the Cochrane Collaboration, an organisation set up to support meta-analysis, which combined results from multiple studies to increase the predictive value of the research. The final issue is that, like anti-HIV drugs, influenza virus has evolved resistance to the drug. The evolution of resistance to the drug suggests it has some impact, but it may need to be used differently to be effective.

Drugs such as oseltamivir and remdesivir might be effective if used as chemical prophylaxis; that is, giving the drugs to people to stop infection before it ever happens. However, they are somewhat toxic, so this course of action could potentially do more damage than doing nothing at all. I spoke to Professor Wendy Barclay (Head of Infectious Disease at Imperial College) about this issue; she said there is a sliding scale of when drug toxicity might be OK. A mild infection, for example a common cold caused by rhinovirus, is preferable to a drug that makes 3% of people vomit. However, taking a drug that prevents a virus that kills 3% of people, like the 1918 influenza, is probably worth the risk.

In the end, we don't know enough about the interface between acute viral infections and the body that leads to disease. It is probably different in different people and so the timing and type of drug treatment will vary. However, this does not take away from the remarkable achievement of antiviral drugs in controlling chronic infections. Thirty years after Freddie Mercury died the terrible tragedy of the HIV epidemic continues, but drugs have miraculously changed the course of viral infection. One of the conditions that Mercury had was bronchial pneumonia and the major cause of this in AIDS is a fungal infection, *P. carinii*. In the next chapter we will look at the impact of fungi on human health, starting with the most dangerous thing to come out of Holland.

CHAPTER 12

Antifungal

Timeline: 3 November 2020. Epsom. Biden wins. Global COVID-19 cases 45,968,799; deaths 1,192,911.

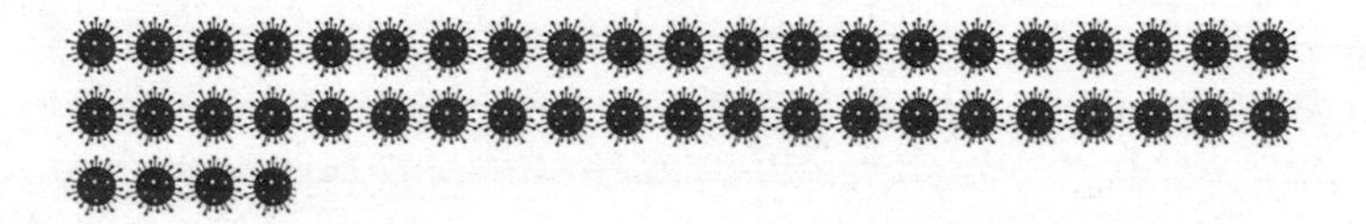

'Dutch tulips from their beds; Flaunted their stately heads.'

James Montgomery

IF I WERE to ask you which Dutch staple posed the greatest threat to human health, Edam cheese, cannabis or tulips, what would you say?

Surprisingly, the answer is tulips. Tulips, as everyone knows, come from Amsterdam and are grown in bulk there: approximately two billion of them every year. This is quite a lot considering the Netherlands is only the 137th largest country in the world by land mass (coming in at 41,256 km^2, just below Denmark and above Switzerland). In order to achieve this horticultural *coup de main*, they need to use intensive growing techniques. One thing cut flowers suffer from is fungal infections. To reduce the impact, antifungal treatments are applied, in huge amounts. The chemicals of choice are the **azoles**: highly potent broad-spectrum antifungals that kill a wide range of the pests that can ruin your beautiful tulips before they reach your vase. Unfortunately, the azoles are also the front-line treatment for a range of fungal pathogens that infect people and their widespread use in horticulture can blunt their value as medicines. But before we turn to antifungal drug resistance, I need to fill in some details about fungal infections, having rather glossed over them in Chapter 2.

When you think of fungi you probably think of mushrooms,* but the fungal kingdom encompasses a broad range of organisms, from the enormous honey fungus (*Armillaria solidipes*), which can form a single organism network covering an area 2.4 miles across, down to the microscopic. As with the bacteria, most members of the fungal kingdom don't care about you or your health. Most fungi are saprophytes and spend their time breaking down dead plant and animal tissue. They are incredibly important in this respect, as they recycle nutrients back into ecosystems.

Fungi are also important in food and drink because the yeast that makes our bread and our beer is a member of the fungal kingdom (*Saccharomyces*). In fact, so much fungi remains after the brewing process that it can then be repurposed to make something even more delicious (or not, depending on your point of view): Marmite. Other fungi make blue cheese blue (*Penicillium roqueforti*), ferment soy sauce and sake (koji mould or *Aspergillus oryzae*) and produce citric acid from corn syrup (*Aspergillus niger*). As well as improving food, fungi can spoil it: the furry mould

* Or pizza, or possibly why the mushroom was invited to the party.

on top of a jam jar or an old piece of bread, which you might potentially scrape off before giving the jam sandwich to your partner (without telling them, obviously), are both fungal. And of course, fungi are important in the control of pathogens: the source of penicillin was a fungi; *Penicillium notatum*.

But there is also a small subset of parasitic fungi and as with the other pathogen classes described in this book fungal pathogens prey on a range of species. Most of my knowledge about the interplay of fungi and their hosts comes from the work done by Professor Matthew Fisher and his team of mycologists (scientists who study fungus, from the Greek *mukes*) at Imperial College London. Mat and his team do amazing work on the interplay between humans, other animals (particularly frogs) and fungi. Each summer they go to a system of remote lakes in the Pyrenees and measure the spread of the fungal pathogen *Batrachochytrium dendrobatidis* and its impact on the frog population there. Interestingly, there are specific words for the spread of pathogens in an animal population – enzootic and epizootic rather than endemic and epidemic, with the Greek derived stem *zoo* (meaning 'of animal') replacing the Greek word *demos* ('of people') in the middle. The same Greek stem *zoo* is used in zoonosis: the transmission of disease between species. It also gives us zoo, short for zoological garden, where one can study the behaviour of animals – specifically whether they enjoy wearing hats, swimming around in tiny pools, pooping on glass screens or eating the arms of stupid people who reach into their cages.

One of the more remarkable examples of a parasitic fungi lives on ants. This pathogen, *Ophiocordyceps unilateralis*, has an amazing life cycle: it infects the carpenter ant (*Camponotus leonardi*), which normally lives in the tree canopies. The carpenter ant lives too high up for the fungi to thrive, but O. *unilateralis* has evolved an ingenious solution. Fungally infected ants fall from the trees to the ground and then migrate to exactly the right height for optimal growth of the fungi. The fungi at this point has complete control of the ant, driving it to lock its jaws onto the leaf and create the perfect vessel for fungal growth, drawing nutrients directly from the plant. Finally, the fungi grows a fruiting body (essentially a mushroom) out of the top of the ant's head, seeding future rounds of infection.

Sadly, human fungal infection does not involve mushrooms growing out of the body: it is closer to a bacterial infection, with single microorganisms colonising tissues. Fungal infections can be external or internal, with a preference for warm, moist tissues. The scale of infection severity moves from the irritating and uncomfortable (athlete's foot, ringworm, thrush) to the opportunistic and deadly (*P. jirovecii*, *Aspergillus* species and *Cryptococcus neoformans*).

Athlete's foot, as you will know if you have ever played too much team sport, been in the armed forces or really anywhere there are naked bodies in close proximity, is a bugger. You can call it *tinea pedis* if you want to be fancy, but I'll still know that you haven't been washing, and drying, your feet properly. Athlete's foot has a characteristic itchy rash and broken skin that won't heal. It belongs to a wider group of fungal skin infections called ringworm, which gets a different Latin name for different parts of the body, so *tinea cruris* or jock itch in the groin, *tinea barbae* (infection of the beard, something Mr Twit no doubt suffered from) and *tinea corporis* (anywhere else on the body). Ringworm is caused by more than forty different types of fungal species called dermatophytes, including *Trichophyton*, *Microsporum* and *Epidermophyton*. The fungi spread tiny hairs or hyphae through the upper layers of the skin, which are less densely packed. These hairs release enzymes to break down keratin, a structural protein that forms part of the skin, hair and nails. The itchiness and redness result from the immune response attacking the infected tissue. Ringworm also causes fungal nail infection – gross, scuzzy, yellow grandad nails which unfortunately won't heal because they have no connection to the immune system, so the fungi can party on uncontained.

Another non-lethal common fungal infection is thrush; a yeast infection caused by members of the *Candida* family, predominantly in the mouth, vagina or penis. It causes white patches on the infected area, which are reminiscent of the white flecks on the thrush bird. It can also lead to a white 'cottage-cheese-like' discharge, which has very little to do with the theme of this book, but if you can't induce a visceral response while writing about infectious disease, when can you? The incidence of *Candida* has increased since the introduction and widespread use of antibiotics. This comes back to the role of the bacterial microbiome in health and disease. The good bacteria in our mouths and genital tracts form a barrier preventing other organisms from colonising those sites. Strains of *Lactobacillus* present in a healthy vaginal microbiome compete with *Candida* for nutrients. The use of probiotics has been proposed as additional therapy for thrush in conjunction with antifungal drugs, but the evidence for any clinical impact of this is weak at best.[1] For both athlete's foot and thrush, the go-to drugs are azoles.

The azole drugs were brought in as a replacement for amphotericin B, the first antifungal drug. Amphotericin B comes from a natural compound discovered in the classical antibiotic screening approach – looking for stuff produced by one microbe that can kill another. In 1958, researchers extracted the drug in New Brunswick, New Jersey (the same town where Waksman was taking credit for his antibiotic discoveries) from a bacteria called *Streptomyces nodosus*, collected from the Orinoco river in Venezuela. Amphotericin B affects the cell membrane (the layer that surrounds them, keeping the cell components inside and the outside world out). The cell membrane is made of lipids and amphotericin B binds to these lipids, punching a hole in the cell membrane which leads to the good stuff leaking out. There is a subtle difference between fungal cell membranes and human cell membranes: fungi use ergosterol and humans use cholesterol. Because amphotericin B has a slight preference for the fungal lipid (ergosterol), it kills the fungal cells more than the human cells. But it is only slightly selective and amphotericin B also affects human cells, leading to severe and often lethal side effects and damaging the kidneys. In spite of its toxicity, as the only available drug it remained in use from 1958 to 1976, until the development of the azoles.

Azoles, as I am sure you know, are five-membered heterocyclic rings containing a nitrogen and one other non-carbon atom . . . nope, me neither. In 1976 Janssen Pharmaceuticals developed ketoconazole, the first antifungal azole drug, which chemically speaking is an azole ring with a whole bunch of other chemistry stuff stuck on the side. Azoles work in a similar way to amphotericin by punching holes in the fungal cell walls, so they are still pretty toxic. Following the development of ketoconazole, a series of other more active, less toxic azole drugs were developed, for example fluconazole. Azole drugs are widely used, having a transformative effect on fungal infections.

The development of drug resistance is an irritant for ringworm and thrush but it is an absolute catastrophe for the big fungal killers – *Pneumocystis jirovecii*, *Cryptococcus neoformans* and *Aspergillus fumigatus* – which contribute to about one million deaths (or a quarter of a COVID-19 pandemic) *every* year. These infections only really take hold in people with compromised immune systems, either through genetics, drug treatments or transplant: the same risk factors as for bacterial infections. The other underlying reason for fungal infection is immune system damage by HIV, which explains why about 40% of pneumocystis pneumonia cases occur in people with AIDS. *Pneumocystis jirovecii* degrades the walls of the lungs, basically suffocating the person from within. Untreated it is a major cause of death in people with AIDS. However, it can be controlled with a combination of azole treatment and steroids. As with the other fungal pathogens, the *P. jirovecii* pathogen is gaining resistance to antifungal azoles.

The source of infection with *P. jirovecii* is not well understood, but it is most likely caught by the inhalation of fungal spores. We breathe in spores all the time. Estimates vary, but measurements suggest that there are 1,000 to 10,000 spores in a cubic metre of air (1,000 litres): since most people inhale half a litre of air that works out as 1–10 spores per breath. Most spores are simply exhaled or caught up in mucus and coughed up later; except in people with damaged lungs or immune systems, in whom they can settle and cause disease. Sometimes fungal disease directly affects the lungs, but at other times the inhaled fungi can spread to other organs. Invasive fungal infection of the brain can be particularly dangerous. Fungal meningitis is most commonly caused by a yeast called *Cryptococcus neoformans*. *C. neoformans* is one of only two fungal pathogens to make it onto Schedule 5 of the Anti-terrorism, Crime and Security Act 2001, a Who's Who of the weird, wonderful and deadly. The other fungi on the list, *Cladophialophora bantiana*, poses considerably less of a threat, having caused forty deaths globally between 2000 and 2010; to put that into context, two people per year die of accidental suffocation in bed in the UK alone. *Cryptococcus* is a much more serious threat, causing something approaching 200,000 deaths per year. It is mostly spread in bird poop and is acquired by inhalation, spreading through the body into the brain and causing swelling, damage and death.

The final fungal pathogens with a significant impact on human health are the *Aspergillus* species. The priest who first identified *Aspergillus* named it after the holy water sprinkler used in churches, the aspergillum; basically a ball on a chain. This genus contains several hundred different members, including *Aspergillus flavus*, which infects cereal grains; *Aspergillus terreus*, a source of statin drugs; and *Aspergillus niger*, the black mould found in damp buildings. The main pathogenic member of the family is *A. fumigatus*, though it probably doesn't mean to be. It is not a fussy eater, spending most of its time in compost heaps breaking down old leaves, grass cuttings and teabags (if my compost heap is anything to go by). However, compost heaps are about the same temperature as the human lung, so given the chance *A. fumigatus* would grow there. *A. fumigatus* has become resistant to the front-line drugs, which brings us back to the Tulips of Doom™. Azoles, because of their extensive use in farming and horticulture, are present in high amounts in compost heaps and wherever there is rotting plant matter. This places a selective pressure on the fungi that live there, basically pre-training *Aspergillus* species to be drug resistant even before they ever reach the human body. This is a similar problem to the use of antibiotics in animal feed and its impact on drug resistance. Only through cross-regulation of both medicine and farming can drug resistance be managed. And it is doubly important in antifungal stewardship, with even fewer drug classes than antibiotics.

The complexities of treating fungal infections are made greater by the genetic proximity of fungi to humans – they are both eukaryotes. The way fungi function is much more similar to us than the bacteria, so finding a unique weak point is that much harder. The final class of pathogen we will look at, the parasites, also have this challenge, but drug treatment is made even harder by their multi-stage life cycles. As with all the other pathogen types, there is both good and bad news: there are highly effective drugs out there, but they are beginning to fail due to overuse and misuse.

CHAPTER 13

Antiparasitic

Timeline: 9 November 2020. At work in London, in a panic still trying to finish first draft. Pfizer/BioNTech release efficacy data. Global COVID-19 cases 49,727,316; deaths 1,248,373.

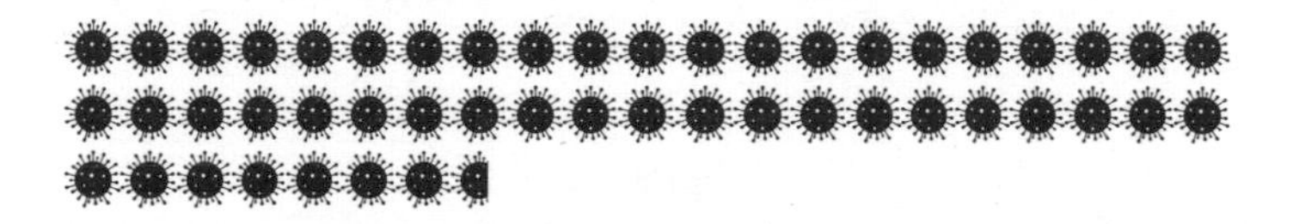

'If you think you are too small to make a difference, you haven't spent a night with a mosquito.'

African proverb

SURPRISINGLY, IN THIS age of designer genomes, personalised medicine and high-throughput chemical screens, one of the most important anti-infective drugs has roots in a traditional Chinese medicine. It turns out that traditional remedies can have merit. Admittedly, not many of them – most of them are largely hokum. It takes a substantial effort to separate the winning drug from the flying squirrel faeces, pangolin scales and other odd ingredients that form part of the traditional Chinese apothecary. In fact, it can take more than two thousand different herbal remedies to find one that works. Fortunately, Tu Youyou and her team undertook this epic task, giving us artemisinin, the go-to standard for malaria treatment. Malaria is an example of the more complex eukaryotic parasites that can infect us. Parasitic pathogens encompass a huge range of different organisms, from single-celled protozoa to enormous tapeworms.

Despite being major causes of death and disease, parasites are often overlooked compared to bacteria and viruses. This is for a range of reasons, not least the complexity of working with these pathogens; for example, to culture the parasite *Trichinella spiralis* requires blending worm-infested rat muscle in a smoothie-maker. Rest assured, the progress in these areas is just as remarkable as it has been with the viruses and bacteria.

MALARIA

Let's start with the most important parasitic infection: malaria. Malaria has been with humans for as long as there have been humans. Indeed it predates humanity – with the recent discovery of malaria parasites trapped in amber from the Paleogene period, thirty million years ago.[1] Both ancient Greek and ancient Chinese texts describe malaria infections. Lord Nelson, Genghis Khan, Dr Livingstone, Mahatma Gandhi and Christopher Columbus all had bouts of the disease. The name derives from bad air – *mal aria* in medieval Italian – because of the prevalence of the disease in swampy areas.

Malaria infection is characterised by fever and tiredness – driven by the destruction of the red blood cells and the resultant anaemia. As with other human infections, for much of malaria's long history treatments were quackery – with the doctors in the Middle Ages resorting to their standard tool kit of vomiting, bloodletting and trepanning.* However, the Quechua people, indigenous to South America, discovered that tree bark from the quina-quina tree could reduce shivering during malaria infection. Agostino Salumbrino, a Jesuit brother and apothecary, observed this use of the bark and shipped some back to Rome as a treatment for malaria. Carolus Linnaeus himself renamed the tree as the cinchona in 1742, after the wife of the Spanish ruler of Peru (the count of Chinchon). Initially it was taken as a tincture of dried bark, dissolved in wine. However, in 1820 French investigators Pelletier and Caventou isolated the bark's active ingredient, quinine, initiating its use as a prophylactic drug. British colonists stationed in India were required to take quinine and discovered that mixing it with sugar and soda water made it slightly more palatable, especially after the addition of gin; pretty much the opposite of today's use of tonic water to take away the taste of the gin!

Malaria and quinine are tangled up in the history of slavery and war. The prevalence of malaria in the American South during the colonial era contributed to the reliance upon African slaves in the Southern Colonies, because they were more resistant than white colonists. Quinine also allowed white Europeans to explore West Africa, which had previously been described as a white man's grave because of the ubiquity of malaria. Cinchona bark became an extremely valuable resource and the government of Peru outlawed the export of the trees. However, in the nineteenth century the

* Though at least they didn't suggest keto diets for viral infections.

Dutch government succeeded in stealing the plants and set up plantations in Indonesia, essentially establishing a monopoly – producing 22 million pounds of cinchona bark by the 1930s.* In the Second World War, these trees fell under the control of the Japanese, reducing access to the drug for the Allied forces. American soldiers in the South Pacific and British soldiers in Burma both suffered very high levels of malaria during the war as a consequence; including my maternal grandfather, Donald Dixon CBE, who caught malaria in Tanzania in 1943 and whose skin turned yellow in response to the quinine substitute, Atabrine, that he used. In 1944, the American chemist and Nobel Prize winner Robert Woodward developed a synthetic method to produce quinine, which alleviated some of the pressure on the troops.

Quinine has several toxic side effects and was replaced by a derivative drug, chloroquine. Hans Andersag discovered chloroquine at Bayer in the 1930s and the Afrika Korps of the Wehrmacht used an analogue of it, sontochin, during the Second World War. Bayer's research came to light after American troops landed in Tunis. Unfortunately, as seen with the antibiotics and antivirals, malaria parasites resistant to chloroquine began to emerge and alternatives were needed. Ann Bishop first demonstrated resistance to chloroquine in Cambridge in the 1920s and '30s. Bishop's research successes came in spite of the barriers put in her way because of her gender; for example, she was forbidden from sitting at the table during departmental tea breaks and instead sat on the first-aid box.

War also played a key part in the discovery of the drug that replaced chloroquine, artemisinin. In 1967, the Chinese

* Colonial theft of trees was rife – see also rubber and tea.

government launched the research programme that discovered artemisinin, project 523, at the behest of Ho Chi Minh, to help protect NVA soldiers against malaria during the Vietnam War. Tu Youyou led the project; curiously, her father named her Youyou after the noise that deer made when they ate the sweet wormwood plant (qinghao) from which artemisinin was derived. Though related, the sweet wormwood plant (*Artemisia annua*) is different to grand wormwood (*Artemisia absinthium*), the source of absinthe. Tu worked in China in the dark days of the Cultural Revolution, when scientists found themselves in one of the nine black categories in Chinese society and as such under considerable scrutiny. In her attempts to develop drugs against malaria, Tu went back to traditional Chinese medicine. Between 1969 and 1971 Tu and her team tested two thousand recipes. She found success in a recipe taken from a book entitled *Emergency Prescriptions Kept Up One's Sleeve* by Ge Hong, which involved soaking qinghao leaves in water. Tu tweaked the extraction method and isolated the active compound, artemisinin, earning herself the Nobel Prize in 2015. The drug is used widely to treat malaria and has saved many millions of lives.

Louis Miller and Xinzhuan Su pieced together the story of the discovery of artemisinin and the role of Tu Youyou from the fragmented records of the Cultural Revolution. If you'd asked eighteen-year-old me, he would have been very sniffy about studying the history of science. Turns out eighteen-year-old me was a bit of a prick. This book would not have been possible without those who pieced together the stories of the discoveries and breakthroughs. So, apologies from eighteen-year-old me and thanks from forty-year-old me to science historians.

Anti-malarial drugs work along similar general principles to other anti-infection drugs. They inhibit unique parts of the pathogen's life cycle. Both quinine and chloroquine inhibit the ability of malaria to process haemoglobin. The malaria parasites infect red blood cells and in so doing absorb haemoglobin; however, too much haemoglobin poisons them unless they can break it down.

PROTOZOAN PARASITES

Malaria is an example of a protozoan, or single-celled, parasite. Another example is giardiasis, caused by a flagellate, *Giardia duodenalis*. Antonie van Leeuwenhoek probably first observed it in his own poo and was clearly a man obsessed with his own bodily functions. *G. duodenalis* colonises the small intestine, attaching to the epithelium and disrupting the uptake of chloride ions, which reduces water absorption. Trust me, *Giardia* sucks. It combines the misery of terrible diarrhoea with eggy burps and farts. I had the misfortune of catching it in 1999. I was 'finding myself' in a very nineties way on an overland bus trip from Egypt to India. The main thing I found in Eastern Turkey was that drinking from the tap can cause you to crap yourself uncontrollably. Luckily, in Iran they had a poorly regulated pharmacy and I got over-the-counter drugs that fixed it. I have no idea what drugs I took, or their mechanism of action, but I certainly reseeded my microbiome. Sadly, I didn't learn my lesson and twenty or so years later, in 2014, I drank from a standpipe tap at the allotment. This time was not quite so bad, so I didn't immediately go to the doctor but instead spent the rest of the summer experiencing mystery bouts of diarrhoea. I became adept at finding public conveniences – let me tell you the UK does not have enough of these – and lost any aversion to pooping in the woods. Finally, I listened to my wife and got diagnosed. The drug I was given was metronidazole.

Metronidazole has a similar history to the antibiotics developed by Waksman. In 1954 researchers at Rhône-Poulenc (now part of Sanofi) were looking for a treatment for *Trichomonas vaginalis*, a protozoan parasite that colonises the vagina. They explored natural compounds produced by the bacteria *Streptomyces*. One compound called azomycin had activity and they synthesised two hundred chemical derivatives, of which metronidazole combined the best profile of activity vs toxicity. It works by disrupting DNA in the pathogen, but to be activated it first needs to undergo a chemical reaction called reduction, which can happen only in organisms that live without oxygen (anaerobes); for example, *G. duodenalis*. One side effect of metronidazole is that it cross-reacts with alcohol to produce another chemical called disulfiram, also known as Antabuse, which is used to treat chronic alcoholism – disulfiram blocks ethanol breakdown and causes instant hangovers. Do not drink booze while on metronidazole!

Malaria and *Giardia* are not the only protozoans that can infect humans. Protozoan parasites often get grouped under the catch-all 'neglected tropical diseases (NTD)'. When you consider that NTD affect one billion people in 149 countries these pathogens deserve a bit less neglect. The WHO has made a list of the top twenty neglected diseases by severity, eleven of which are caused by parasites, and three of these are protozoan: Chagas disease (American trypanosome), sleeping sickness (African trypanosome) and leishmaniasis.

Chagas disease, caused by *Trypanosoma cruzi*, is transmitted by the kissing bug (triatomine). It leads to fever and heart disease and drove Charles Darwin to write: 'But I am very poorly today and very stupid and hate everybody and everything.'* *T. cruzi* infects nearly seven million people, mostly in Latin America.

The other major trypanosome to infect people is *Trypanosoma brucei*, the cause of sleeping sickness. The tsetse fly transmits *T. brucei*, causing cerebral infections leading to the difficulty in sleeping from which the disease takes its name. The WHO has set out to eradicate sleeping sickness. In the year 2000, there were 26,550 cases of sleeping sickness; by 2018 this had been reduced to only 977. In August 2020, despite everything else that happened during the COVID-19 pandemic, through a combination of drugs and insect control, Togo was the first African country to eradicate sleeping sickness. The WHO projects complete prevention of the disease by the end of 2020 and complete eradication by 2030.

Another parasitic disease with a considerable impact is leishmaniasis, caused by various species of the protozoan *Leishmania*. Sandflies spread *Leishmania*, particularly in the tropics, where it affects six million people, with nearly one million new cases a year. In honour of the discovery of *Leishmania* by William Leishman (in the UK) and Charles Donovan (in India), Ronald Ross, the doctor who linked the plasmodium parasite with malaria via the medium of poetry, named one of the strains *Leishmania donovani*. As with malaria, *Leishmania* has a complex multi-stage life cycle, cycling back and forth between flies and mammals:

* Which pretty much summarised 2020 for me.

the parasite uses the host immune cells to mature and proliferate. There are three forms of the disease: cutaneous, characterised by a large open sore; mucocutaneous, characterised by skin and mucosal ulcers; and visceral, where internal organs are infected. Visceral leishmaniasis, also called kala-azar or black fever, is fatal if left untreated. The magnificently titled Rai Bahadur Sir Upendranath Brahmachari FRSM FRS developed one of the earliest treatments, a drug called urea stibamine that incorporated the heavy metal antimony, selectively poisoning the parasite rather than the person.

DRUG MONEY

Sadly, the neglected tropical diseases reflect the economics of drug development. Getting new drugs to market costs a lot of money. The most commonly quoted figure for the cost of bringing a new drug to market is one billion dollars; this includes the dead ends, failures, trials, volunteers and manufacturing. This massive wedge of cash has to come from somewhere and in a capitalist economy that is from sales. Companies need to make a profit to survive and drugs are no different to cars, Coca-Cola or Coco Pops. Without coming across too Alan Sugar, anyone can sell a £10 note for £5. As with the lack of novel antibiotics, the lack of economic return has led to a lack of investment in many of the diseases described in this chapter. There is a pressing clinical need for new antiparasitic drugs. Between 1975 and 1999 there were only thirteen new antiparasitic drugs developed. One alternative funding approach is philanthropy and the antiparasitic field has been reinvigorated by investment from the Gates Foundation.

The lack of investment in neglected diseases by commercially driven pharmaceutical companies is not necessarily unethical in and of itself – though it does perpetuate an uneven world. But there are instances where approaches taken by companies to sell antiparasitic drugs have been clearly unethical. A well-known example of this is the sorry tale of Turing Pharmaceuticals, Martin Shkreli and Daraprim. Daraprim is a treatment for the protozoan parasite *Toxoplasma gondii*, which causes the disease toxoplasmosis. *T. gondii* infects up to half of the world's population. It can be caught from two sources: infected meat or cat crap. The normal life cycle of the parasite is that infected mice get eaten by cats, the cats then excrete infected poop, which gets eaten by more mice, which are then eaten by new cats. Curiously, infected mice become less fearful of cats, increasing the likelihood of parasite spread when they are inevitably eaten. Humans represent a dead end for the parasite – unless, of course, you are a crazy cat lady and get eaten by your wayward pets. Most people are infected asymptomatically, though it can be dangerous for unborn babies.

Toxoplasmosis poses a substantial problem in people with compromised immune systems, most notably people living with HIV/AIDS; this is reflected in the novel *Trainspotting*, with one of the characters, Tommy, dying of it.[2] Daraprim can be used to control HIV-associated toxoplasmosis, which is where our villain, Martin Shkreli, steps in.* Shkreli, a hedge fund manager, made his money in a

* Martin is admittedly not your traditional villain's name. It is more reminiscent of a retired auditor living at number 6 Privet Drive, Little Whinging.

variety of ways too complex for an immunologist to understand, but sufficiently bad to land him in prison for seven years. Before being convicted, Shkreli set up Turing Pharmaceuticals. Unlike most pharmaceutical companies, Turing didn't develop anything new; it merely bought licences for other drugs and then repacked, repriced and sold them on. In 2015, Turing increased the cost of the drug from US$13.50 to US$750 per pill overnight – a 5,500% hike. This was, unsurprisingly, unpopular leading him to receive the title of 'Most Hated Man in America' (for a brief while, because competition for that slot is tight and currently occupied). The only good news is that his move spurred another company, Imprimis/Harrow Health, to make a generic version of the drug for $1 a dose. The move by Shkreli/Turing was possible because only one company manufactured the drug, giving them a de facto monopoly.

Drug patents last twenty years from the moment the company registers them. This number is a compromise between the needs of the company and the needs of the patients. While a drug remains on patent, the company will effectively own a monopoly and can charge, within reason, what they want for it. This money will then seed future drug development (and pay the shareholders). The companies normally patent their compounds before they know for certain they will work – because clinical trials are expensive and finding out that your drug works but then not recouping the substantial investment would not be a good business tactic, as happened to Bayer with Prontosil. But registering a drug before it is ready to sell eats into the time to recover the cost, which can further increase the price during the monopoly period. When a drug comes off patent, anyone can then manufacture it – these are called generics and are normally much cheaper. For many drugs, this will then lead to multiple companies producing the same drug, driving the price down in classic supply and demand economics. For example, most of the drugs you buy over the counter at a pharmacist will be generics: it explains why you can buy a packet of sixteen 200 mg ibuprofen tablets for 55p compared to a packet of sixteen 200 mg Nurofen™ for £1.89 (prices from Tesco.com on 2 February 2021).

Generics have been incredibly important in the control of HIV, enabling charities and governments to get cheap drugs to the people who need them the most. However, it hasn't always been straightforward. In 2000, the drugs were available but weren't getting to the people who needed them the most, because they cost somewhere between $10,000 and $15,000 per patient per year (to put that in context the average healthcare spend per capita in Africa is $80). A set of World Trade Organization agreements called TRIPS, put in place to protect patents, contributed to the problem. TRIPS awarded a disproportionate benefit to the high-income countries, who generate more of the patents and have less of the disease. Prior to TRIPS, patents on medicines had not been respected in all countries, particularly when the patents were not in the public interest. In 1997 the post-apartheid government of South Africa made laws allowing them to produce generic anti-retroviral drugs. The battle between the Nobel Peace Prize-winning Nelson Mandela and Big Pharma certainly drew attention to what had until that point in time been a somewhat obscure piece of international patent law. Brazil and India also took on the patent holders, producing generic versions of HIV drugs. A combination of the court case in South Africa and pressure from civil society groups led to a rethink of the

TRIPS agreement in the context of low-income countries, allowing countries to 'promote access to medicines for all'. There needs to be a major rethink about the difference between the cost of drugs and their actual value and I will return to this in the epilogue.

Even with generic antiparasitic drugs, there are relatively few options for the treatment of protozoan parasites. The situation is even worse for people infected with the other type of eukaryotic parasite: worms. Worms infect far more people than protozoan parasites (even taking malaria into account).

WAITING FOR THE WORMS

While the single-celled parasitic protozoans are complex in comparison to viruses or bacteria, they pale into insignificance compared to the worms. The longest creature alive is a parasitic worm – those found inside the guts of whales can be over 100 feet long, the length of three London buses parked back to back – longer than the whales themselves!* Humans can also be infected by tapeworms, which can live for an incredibly long time inside the host (up to thirty years), allegedly growing to eighty feet long; though I struggled to find actual scientific literature to back up this number.

Eight of the eleven official neglected tropical diseases are caused by worms. A substantial proportion of the world's population is infected with one of three worms, described as helminths – ascaris (807–1,121 million), whipworm (604–795 million) and hookworm (576–740

* Told you it was a useful unit of measurement.

million); though some people will have multiple infections. The infections are concentrated in the poorest half of the world's population. While not acutely detrimental to health, helminth infections can cause stunting, reduced physical fitness and poor performance in school, leading to a negative cycle of poverty. Helminths are spread through contaminated soil. Ascaris and whipworm are transmitted as eggs and acquired from unwashed vegetables or hands. Hookworm is acquired as larvae, which bury themselves into the skin on the soles of bare feet (shudder). Parasitic worms influence the immune system, shaping the responses to other pathogens to hide themselves from the body. The use of worms as a treatment has been explored in the context of the hygiene hypothesis; investigating whether individuals with ascaris infection are less likely to develop allergies. My wife's PhD involved isolating proteins from parasitic worms as novel anti-inflammatory drugs.[3]

Worm infections can be treated with two drugs – praziquantel and albendazole. Both praziquantel and albendazole were initially developed for treating farm animals and were subsequently tested in people. Which sort of makes sense. The bar for animal drugs is considerably lower than human drugs and there are also more animals, which makes testing a lot easier. But it also reflects the problem with prioritising profit over human life. Many antiparasitic drugs started life targeting other conditions but were then repurposed, including eflornithine (trypanosomes) and miltefosine (leishmania), both initially anti-cancer agents, and praziquantel, a possible tranquiliser. The overlap between antiparasitic drugs and other targets is not overly surprising. Eukaryotic parasites have many more similarities to human cells than bacteria, making it much harder to find unique aspects of their life cycle to target.

EAT MY WORM

The complexity of the parasite–human interaction necessitates further investigation in order to develop new vaccines and drugs. Deliberate human infection studies can help dissect the response; a similar approach to that used by Jenner for smallpox and the Common Cold Unit for influenza. Aside from malaria, less work has been done on human parasitic challenge models using parasites. However, in the last few years infection models for other parasitic infections have been explored, including *Leishmania major*, by Professor Paul Kaye in York. Another model being developed is for schistosomiasis, by Dr Meta Roestenberg at Leiden University in the Netherlands.

Schistosomiasis is also known as bilharzia after the German doctor Theodor Bilharz, who first identified it in Egypt in the 1850s. Bilharz has an odd legacy; on the plus side he has a crater on the moon named after him, but on the minus side he is buried side by side with a Nazi war criminal. As with other parasites, *Schistosoma* worms have a complex multi-part life cycle. Freshwater snails spread the infection by releasing larvae into water, which burrow into human skin. These larvae then mature, spread around the body and ultimately release more eggs via the faeces into water, which hatch and then infect more snails. The disease is characterised by an itch at the point of entry, acute fever and then a range of symptoms depending upon where the worms end up in the body, with the potential to damage the guts, brains and liver. There is also a genital manifestation leading to lumpy semen (which can be confused with an STI) – discovered by 'the intrepid Claude Barlow who deliberately infected himself with cercariae in 1944 and then observed the . . . appearance of schistosome ova in his own semen'[4] – clearly a man taken by van Leeuwenhoek's self-abuse.

Approximately 220 million people worldwide are affected by schistosomiasis, especially those living in close proximity to contaminated water. Large-scale irrigation projects set up in the 1950s and '60s made the problem worse: the dams disrupted the life cycle of predatory prawns that hunt and eat the snails. Recent studies reintroducing the prawns have had a significant benefit by both reducing snails and providing a source of food and income to local communities. To understand immune responses to the parasite, Dr Roestenberg has developed a human challenge model; in order to do these studies she has to keep a colony of snails that are infected with the male worms.[5*]

As with other parasitic diseases, schistosomiasis has been somewhat neglected in the past, but there are grounds for hope. The disease is slowly being brought under control. By 2018, the schistosomiasis control initiative established at Imperial College London with support from the Gates Foundation had delivered 200 million drug treatments against parasitic worms. The costs per person are tiny – £0.33 per child per year – and the results are spectacular. Reducing the burden of infection means the food the children eat feeds them and not their worms, giving them fuller, more productive lives.

* Insert your own joke here about the pace of the research.

Thanks to Tu Youyou, Trudy Elion, Selman Waksman and all the researchers named and unnamed we have tools to fight infectious disease by fighting off the pathogen, either through the use of drugs or training the immune system with vaccines. But these tools are only effective if you use them properly (or even at all); an important piece of the puzzle is to try and understand why people don't and the bad science that underpins their decisions.

CHAPTER 14

Bad Medicine

Timeline: 23 November 2020. London still. Oxford/Astra release efficacy data. Global COVID-19 cases 57,882,183; deaths 1,377,395.

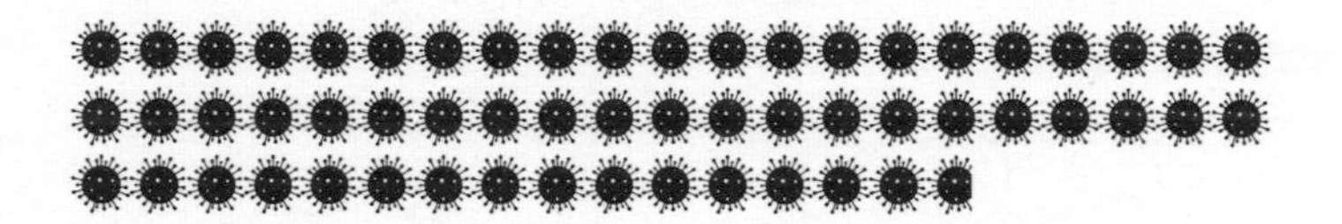

'Beware of false knowledge, it is more dangerous than ignorance.'

George Bernard Shaw

LET'S BE CLEAR from the start, I do not believe the conspiracy theories. I do trust the scientists (most of them). If you have read this far and were hoping for a huge plot twist at the end, you are going to be disappointed. The SARS-CoV-2 virus caused COVID-19, not 5G, not asteroids, not the Chinese government; just a virus uniquely adapted to our interconnected world. It might seem odd to have got fourteen chapters into a book about pathogens and still need to write that. But it reflects the sorry state of affairs surrounding the pandemic in some quarters. COVID-19 provides a handy primer on misinformation, critical thinking and how the interface of infection with society means scientists need to deliver their messages in a clear, unambiguous way to save lives.

So why did COVID-19 trigger a deluge of conspiracies? As a rapidly evolving pandemic, it ticked three critical boxes that amplified the misinformation: it was scary, it was unknown and it was disruptive.

THE FEAR FACTOR

A global pandemic of a new virus is scary. I can vividly remember the HIV/AIDS outbreak in the early 1980s; HIV spread invisibly – infected individuals were asymptomatic in the early stages and it was nearly always fatal (thankfully drugs have changed this). Think back to spring 2020 and COVID-19, when we were living through an outbreak of an unknown but rapidly spreading respiratory virus. In the absence of concrete information, we sometimes filled the gaps from fiction – I would recommend reading *Station Eleven*,[1] which describes the consequences of a catastrophic influenza outbreak, but don't read it during an actual pandemic; it will do nothing for your mental wellbeing.

At the beginning of the outbreak comparison to other pandemics provided the main, but not necessarily more reassuring, source of information. The 2003 SARS outbreak (caused by a similar virus with a high fatality rate) and the 1918 influenza pandemic (because of the speed at which it spread worldwide) served as reference points – combining them made for a pretty terrifying scenario. Luckily, COVID-19 hasn't been the worst case – trust me, there are worse outcomes; for example, if the vaccines had no impact (airborne HIV would be much, much worse). But the virus did spread quickly and while not at 15% like SARS-CoV-1, 1% is a very high case fatality rate, especially given the number of cases. And it was scary, especially in March 2020, and remains so for those who get hospitalised not knowing whether they will ever come out.

THE GREAT UNKNOWN

The absence of actual information contributed to the abundance of misinformation about COVID-19. SARS-CoV-2 is a total newcomer: at the beginning of 2020 we knew next to nothing about the virus. Coronaviruses were something of a backwater relative to other viruses. I remember meeting a coronavirus researcher at a conference in 2018 and thinking why are they working on that dead end virus? Which goes to show how ineffective my science radar is!

In the period from 2009 to 2019 there were 6,982 scientific publications on coronaviruses, which sounds a lot, but there were 160,849 on HIV and 66,923 on influenza in the same ten-year period. We certainly knew more in spring 2021 than we did in spring 2020 when I started the book: more papers were published on coronaviruses in 2020 (66,295 and counting) than in all the preceding years since their discovery in 1949. But mysteries remain, even within the bounds of knowledge about other coronaviruses. In some respects SARS-CoV-2 behaves differently, but in others it is the same. Understanding the similarities is as important as the exceptions.

The absence of scientific data allows other ideas the space to grow. There was such an appetite for information and, of course, fear sells copy (or clicks). This appetite for news wasn't always sated by the scientists, who are prone to caution in the face of the unknown – and rightly so, as there is so much we do not know. Scientists are trained to question and are often reluctant to make definitive statements because of the limits of their knowledge. It is considerably easier to be a charlatan, making claims based on nothing. The claims can be stated with absolute certainty because they needn't bear up to actual scrutiny – they are merely bombast. As Mark Twain didn't say 'a lie can travel halfway around the world while the truth is still putting on its shoes'.[2*]

DISRUPTION

The final contributor to disinformation was disruption. The disruption had mild aspects – having to wear a face mask, having to wash your hands or not being able to see your mates in the pub. It also had more significant consequences – loss of jobs and livelihoods and not being able to pay the rent or feed the kids. In the face of an invisible agent that only infected a relatively small percentage of the population at any one time, the response may have felt disproportionate; especially if you hadn't been infected.

Fear, ignorance and disruption led people to look for information to support their point of view, with a confirmation bias in how they found the information. I confess my guilt in this when it came to masks. As a glasses wearer I got to choose between not seeing because my glasses were fogged up or not seeing because I removed my glasses. There

* This quote has been misattributed to Twain, but it probably originated with Jonathan Swift.

was (from my selective reading) a lack of clinically controlled evidence that masks work, which justified my reluctance to wear them before they became compulsory. However, my friends found themselves pretty convinced by a lady with a jaunty beret and a cartoon about wetting yourself, which goes to show public health messaging comes in a variety of forms. The principal argument for masks seemed to be that absence of evidence is not the same as evidence of absence – with the analogy that you don't need a clinical trial to show a parachute working. It was all a bit circular: if people don't wear masks or use them properly, they won't work. In the end, being a rule-follower I wore them where required on the assumption that they did not actively cause harm and probably did some good. But this goes to show you can relatively easily find information that supports your point of view – with a difference between knowledge and wisdom.

In 2020 you didn't have to look far to find the contrarian view. Some of it was brilliant nonsense – for example, the rumour of Wembley Stadium's conversion into a giant lasagne oven. Yet some of it was plausible – for a long time I believed the rumour that leaving hand sanitiser in your car on a hot day could cause the car to catch fire. It turns out that your car would need to be at 350°C for this to happen – 70° hotter than the ignition temperature of petrol and only slightly cooler than the surface of Venus. I would recommend checking in at fullfact.org, who do an excellent job of debunking rubbish.

Unfortunately, some of the misinformation was pumped out by 'scientists', or at least people with a plausible-looking science background to give them some gravitas. Much of the debate (and when I say debate, I mean angry shouting matches reduced to 280 characters) happened on Twitter and other (anti-)social media outlets, which went through cycles of utility and rage. Doom scrolling through pages and pages of science (and politics) became one of the most popular pastimes of 2020, alongside sourdough baking, binge-watching box sets and getting a dog. I'm not racing to name and shame all of the semi-legitimate hucksters who pimped COVID-19 misinformation, but many of them should know better.

FLAT EARTH

COVID-19 isn't the only situation to generate crazy rumours. Some of my favourite conspiracies are that chemicals in the water turn frogs gay, that a clone named Melissa replaced the pop singer Avril Lavigne (because why and who cares, apart from Avril presumably) and that Pope Sylvester added 297 years to the calendar so that he could be pope in the year 1000. But if I had to choose my *ne plus ultra* of conspiracy theories it would be that the earth is flat. I genuinely cannot understand why anyone would choose to believe this when it is so palpably untrue.

Conspiracy theories share some key features[3] linked to their main driver that the 'official account' is wrong: anyone publishing information in line with the 'official version' is in on the act and as such has nefarious intent; there is a vague suspicion that 'something's wrong' and that there are victims of 'the man'; and finally, while the theory can happily incorporate internally contradictory beliefs, it is immune to actual contrary evidence – it can never be disproved. This lack of falsifiability acts as an important tell of a conspiracy theory. A key tenet of science is that it can potentially be proved wrong – the lack of a way to test an idea makes it less, not more, robust.

A lot of psychological research has been done into why people accept conspiracy theories.[4] One idea is minority influence: the people pushing the argument – the flat-earthers and anti-vaxxers – believe it so strongly that they come across as coherent about an idea that is patently bobbins. Another factor is the relative youth of science and scientific thinking in terms of how humans approach problems. Our brains are hard-wired to look for patterns – being able to determine a tiger from a stripy leaf helped our ancestors survive long enough to breed. This genetic predisposition to detect patterns means we find it hard to accept the fact that life, as Thomas Hobbes put it, is 'nasty, brutish and short'.

As Ford Prefect wrote, the flat earth is mostly harmless. But flat-earthing may be a gateway conspiracy to other more harmful ideas. Research suggests that people who buy into one conspiracy theory are more likely to believe another.[5] Some ideas are dangerous, particularly those relating to infectious diseases, when they stop people from doing things in their best interests – not wearing masks, favouring untested drugs over working ones, not taking vaccines and even denying the existence of the virus. I naively imagined that anti-vax would be the most dangerous COVID-19 conspiracy; after all, there has been anti-vaccine sentiment for as long as there have been vaccines. But I was surprised by the force of the small band of COVID-19 denialists, because it is SO CLEARLY REAL, YOU NUMBSKULLS. These monsters have done unconscionable damage: both through subtle influence, where the swaying of popular sentiment has delayed lockdowns leading to more deaths, and the more direct – with heartbreaking videos of mobs shouting at NHS staff and misguided relatives trying to take loved ones off ventilators and demanding they be placed on a regimen of vitamin C and zinc (and Wuffle-Dust, presumably). I cannot put my finger on why people start these rumours, but I can see how they gain traction by flooding the market with misinformation, some of which trickles through.

CRITICAL THINKING

Critical thinking is our pathway through all this hokum. Critical thinking – the analysis of (actual) facts to form a judgement – requires evaluating the quality of the information: not all sources are created equal. You need to distinguish legitimate ideas from fraudulent ones by choosing the right sources of information, from the right people. I would highly recommend *The Demon-Haunted World* by the astrophysicist Carl Sagan as a primer to critical thinking.[6*] But here are some brief thoughts relating to infection:

Choose the right information

Discerning the value of information requires an investment of time. The internet makes getting information easier, but the net is egalitarian – it is just as easy to find bad information as good. The way in which search engines return answers to our questions has historically been a bit circular, with search algorithms skewing towards our previous searches. And there is a suggestion that some algorithms

* Sagan was exceptional. Amongst other things, he performed some of the earliest work demonstrating that life could emerge from a soup of chemicals. He also curated the first record to be sent into space.

tend to the extremes as a way to generate traffic. So, if you start down a bad path it can be difficult to return. However, during the COVID-19 pandemic some of the social media platforms previously associated with the spread of disinformation have made more active efforts to highlight verified sources.

So how to find the right information? The first rule of thumb is use trusted sources – in science, if it hasn't been peer reviewed it isn't worth shit. But what does peer review mean? To understand it, we need to look under the hood of how scientists share their ideas. By and large, scientists communicate their findings in the form of scientific papers – short(ish) reports of research published in scientific magazines.* There are big generalist journals that also do some scientific news reporting, such as *Science* and *Nature*, and there are also specialist ones which sound mostly like they belong on the obscure magazines round of *Have I Got News for You*. In theory, you can't just publish any old rubbish in these journals. Papers face two levels of scrutiny – the editor and the reviewer(s). Unlike newspaper editors, scientific editors tend not to commission stories: instead, they choose which of the reports they receive to publish. The decision of what to publish is a bit of a mysterious black box process and is based on an X-factor quality 'impact'. Impact is basically how many people will read your work, so the goal of a scientific editor is not completely different from a newspaper editor – getting people to read what's published. At its best scientific editing acts as curation, by finding and signposting the best work, and at its worst it acts as gatekeeping, which limits the development

* We call these journals to make them sound more grown-up.

of new ideas; it is normally somewhere between the two.

If the report passes the editorial threshold for interest/ relevance they will pass it on to subject experts for a more detailed review: called peer review. The experts will read the report thoroughly to check that it is rigorous and makes sense and that the conclusions are supported. The reviewers will make comments to 'improve' the report: some of which are sensible, some just daft. Then the author must either do what the reviewer asks or make a case to explain why not. This process can take several iterations and some papers take longer than others; my record is thirteen months of review for a piece of work that only took six months to perform. These reports are called primary research papers. If particularly exciting, they may generate commentaries or over-the-top press releases from the institution that performed the work, some of which will filter into the real world outside of academia. Unless you work as a scientist, going right back to the primary research paper is probably a bit over the top. Instead, when the story is reported in the mainstream media, look to see whether it has been published in a scientific journal – valid news stories should tell you where they got their information from.

Primary, published research is more reliable than hearsay, opinion or the lunatic rantings of 'sceptics' on YouTube. But unfortunately there are publications and there are *publications*; some journals apply a greater degree of scrutiny to what they publish. Just because it is published somewhere, doesn't make it true. Knowing whether the science breakthrough reported in the news has come from a respectable journal is tricky even for specialists working in the field. There are certainly bad journals and fortunately an American university librarian called Jeffrey Beall has compiled a list (https://beallslist.net/) of predatory journals which publish data with little or no scrutiny. Predatory journals basically scam money from unwary scientists.

Be wary also of claims relating to pre-prints: scientific reports yet to be peer reviewed. Pre-prints in the infection field are mainly found on BioRχiv and MedRχiv (pronounced Bio-Archive and Med-Archive respectively, because the Greek letter χ is Chi). They are a useful way for scientists to communicate with each other because of their immediacy, but they lack some of the scrutiny of peer-reviewed papers. Although 2020 was very much the year of the pre-print in biomedical science, I am still in two minds about their value.* In a rapidly moving pandemic they worked to get the information out faster, but they should be treated with a pinch of salt, especially when accompanied by a hyperbolic press release. The alleged detection of SARS-CoV-2 RNA in wastewater before the first recorded cases (described in Chapter 7) is a good example of science by press release; where scientists present their ideas to journalists without going through the normal rigorous scrutiny.

Poor-quality research reported through the press with no scrutiny can be dangerous, especially when it influences policy decisions. Inexplicably, many people decided that an anti-malarial drug (hydroxychloroquine) would work wonders on a virus – SARS-CoV-2 – in spite of clear clinical evidence it had no effect.[7] It wasn't the only anti-parasitic drug proposed against COVID-19 because ivermectin, an anti-worm drug, also gained an unhealthy following. While these drugs *can* work in a lab when used at ridiculously high

* In an entirely unscientific poll of my peers on Twitter, out of thirty-eight respondents 86.8% said they are good and 13.2% bad.

levels unattainable in the human body (the patient would die from the drug first), there has been no evidence they work in the clinic. Advocating for the misuse of drugs is bad because taking them unnecessarily will cause some harm (hydroxychloroquine can cause heart problems) and it may also reduce their effectiveness against their actual targets by increasing drug resistance.

FRAUD

There are, of course, times when the science is wrong. Actually, there are a lot of times when the science is wrong; that's kind of the point. Science evolves constantly. The ideas we had yesterday will be shaped by what we know today to become the ideas of tomorrow. It is an evolution of ideas building one upon another. It is fine and normal when scientists make legitimate mistakes of interpretation based on the data and the theoretical framework available. For example, it is not unreasonable to conclude that influenza was caused by a bacteria – if you didn't know there were viruses and you could reproducibly isolate the *same* bacteria from people with the *same* symptoms.

Self-correcting science is a good sign that the process works. But another kind of wrong exists – when the authors make up the data. You might wonder why they would do this. The main driver is the connection between scientific career progression and the volume and prestige of your publications. There are huge incentives to publish bigger and better stories. If you publish a big shiny scientific paper in a big glossy journal you get more money to continue your research. This enables you to publish more shiny science in a virtuous loop. As with most human endeavours, there is only a limited amount of fame, fortune and funding and it is spread unevenly. Much to my father's disappointment, science is not a universal love-in. There are just as many egomaniacs in research as any other field with a zero-sum model (luckily there are also a lot of great people, with whom you can slag off the plonkers). The incentives to win, therefore, can pressurise people into making bad decisions. Some scientists become so driven to make a big breakthrough in their fields that they simply make up the data.

Why are the fake papers so damaging? Firstly, false papers skew the record, sending researchers down the wrong path. Science builds upon the foundation of everyone else's work. If some of that foundation is wilfully wrong, then the whole edifice wobbles. Following a bad trail laid by a fraudster is a waste of money and time; and I ask you to spare a moment here for researchers who follow these fraudulent paths and waste their scientific careers because some other fucker lied.

Bad papers also poison the well. More tragically than misleading unwary researchers, they can offer false hope to people suffering with a condition – as seen with the retracted XMRV-1 story in chronic fatigue syndrome. People will also use the bad scientific papers as a justification for their actions. The retracted MMR paper has led to a dramatic increase in measles cases: in the USA, there were more cases of measles in the first month of 2015 than in any year since the vaccine came out.

Ideally, false data would be picked up at peer review but unfortunately this doesn't always happen, for a number of reasons: the assumption is that most scientists are honest, so fraud is not always at the forefront of the reviewers' minds; the reviewer might not be quite expert enough to pick up on the nuance; or they might be so in awe of the author that they don't apply as much scrutiny as they should. In defence of the reviewing community, peer review is done as a voluntary service on top of everything else the scientists are trying to do and occasionally it can be rushed. Peer review is a bit like democracy: it is the least bad system we have.

While the buck stops with the individuals who write the false paper in the first place, the publishers of the journals do have a duty to try and stop these papers from ever being published. The main approach has been to get authors to sign pages and pages of declarations that they have not committed fraud; which to me feels as effective as the declaration on the US visa waiver where you tick a box stating you are not a Nazi war criminal.

Mostly the bad papers will then get identified and removed from the scientific record – in a process called retraction. Retraction of your paper is bad and normally it is career-stoppingly so. Some scientists, though, are serial fraudsters – Yoshitaka Fujii, a Japanese anaesthesiologist, has had 183 retractions, a staggering number. It is impossible to know the motivation of Andrew Wakefield and the MMR story – whether he thought he was correct or was deliberately fraudulent – but it is of note that he has only had two papers retracted, suggesting compounded error rather than malign intent.

Eventually, the wider community of scientists will read the paper and pick up on the errors. There are even professional fraud spotters – Elisabeth Bik, a microbiologist, works full-time investigating image fraud in papers (where a picture has been manipulated to give the result wanted). And two scientific editors, Adam Marcus and Ivan Oransky, founded a website called Retraction Watch in 2010 to shine a light on this murky field; it is worth visiting to understand the range of scientific fraud possible.

POWER TO THE PEOPLE

So now that you know how to find the right information, how do you find the right people to trust? The first and foremost rule is that when looking for advice about scientific issues you need to listen to scientists. Ian Brown, Van Morrison and Right Said Fred may have made good music (well at least Ian Brown did), but they shouldn't be your go-to source for information about a complex, rapidly evolving pandemic. But 'listen to scientists' only narrows it down a bit: which scientist should you trust? Don't just back an idea because someone is going against the current thinking. However tempting it may be to back the person 'fighting the machine', they are often tilting at windmills. The dissenting scientist voice makes a compelling narrative, so much so it has drifted into popular culture. When they aren't being actively evil, scientists in films are often portrayed as plucky loners battling against the odds; for example, Professor Selvig (*Thor*), Bruce Banner (*Hulk*) and Hans Zarkov (*Flash Gordon*).

Of course, there is a small handful of scientists who did successfully buck the establishment view at the time: van Leeuwenhoek's demonstration that there was microscopic life; Ignaz Semmelweis's demonstration that handwashing prevented infection; and Barry Marshall's demonstration that *Helicobacter* causes ulcers. But these are outliers: science is mainly evolution, not revolution. The people with the biggest impact on our treatment of infectious disease either worked unrecognised in the background like Hilleman or built directly on the work of others, such as Waksman. Being an outlier doesn't automatically make you right; more often than not it makes you wrong. For every one visionary there are a thousand cranks.

COVID-19 conveniently gives us a chance to test our critical thinking. There were some false ideas that only required child-level critical thinking – ingesting bleach, mobile phone masts and the belief that the virus didn't exist or cause disease.

But we also saw ideas that were wrong but sounded right and required careful unpicking; for example, herd immunity. Various scientists hypothesised that maybe there had been enough circulating, asymptomatic virus cases that the outbreak would just peter out (there would be no second wave). A subtle alternative was that you could let the young and healthy become infected while isolating the sick and vulnerable. Herd immunity acted as a siren's call, because no one wanted to be in lockdown and if bona fide scientists from Oxford and Harvard were saying it would all just vanish, maybe it would. The method of presentation of the herd immunity hypothesis to the larger world really rankled – instead of publication as a peer-reviewed paper, it appeared as a grandiose declaration which was sponsored by a political think tank, toasted with champagne and then made available for people to sign online.*

But as the Chief Medical Officer, Chris Whitty, pointed out to the Science and Technology Select Committee, herd

* Though I did enjoy people adding fake signatures, such as Dr Jonny Bananas, Dr Person Fakename and Professor Ita Role (pudding and dessert expert), reminding me of happier days when I added Dr Hugh Jars to various university sign-up lists.

immunity was ethically, operationally and scientifically flawed. Ethically flawed because even in younger people there is both a risk of death and of extended after-effects known as Long COVID-19, which is associated with a range of symptoms including extreme tiredness, shortness of breath and brain fog. Operationally flawed because if the old and vulnerable were to be completely isolated who would look after them, how would they see their families and how would they live any kind of life? And scientifically flawed because we have never seen herd immunity generated to any other infection ever – hence the endless circulating influenza, TB, malaria and RSV. The massive resurgence of COVID-19 in the autumn of 2020 in the UK and other countries demonstrated that the core idea was wrong: there weren't enough immune people to prevent another outbreak. In spite of the clear flaws in the argument, the herd immunity declaration provided a nucleation point for misunderstanding and dissent.

'FOLLOWING THE SCIENCE'

As seen in the case of herd immunity, science and politics are intertwined. The link between science and politics never goes away. Some of the time it is quite subtle, for example in the prioritisation of research funding. In my ideal world, I could potter around investigating interesting things and someone else would pay for it because doing science is a 'good thing'. Sadly for me, this isn't the case: most funding comes from the taxpayer via the government, which requires understanding by politicians and the public about the need for research and evidence by the researchers of the value provided by the work they do. The value provided by research comes in various forms: it can be indirect through training, esoteric through knowledge generation or translational through drug development. The role of politics in research can also be much more direct when it does not meet the administration's agenda; for example, HIV research in the Reagan era. Various other blocks have been put in place, including those on the development of stem cells from aborted foetuses or on gain of function influenza viruses.

The science–politics interface is also critical in the public acceptance of new technologies, including drugs such as Salvarsan and PrEP as approaches to tackle STIs; vaccines ('nuff said); GM crops; and even energy-saving light bulbs.

And then there are times when science becomes a political football, with clear sides; for example, the climate change debate. Interestingly, the COVID-19 pandemic, while divisive, did not split down standard party lines (at least in the UK): there were denialists on both sides. COVID-19 really put science front and centre in the public eye. This has mostly been good because there should be informed science-led debate behind policy decisions relating to scientific issues. The interface between politics and science mostly involves persuading the public about the merits of the argument. This is particularly important when the decisions involve discomfort, disruption and distancing. But it can also play a role in the spending of taxpayers' money. In the absence of a catastrophic viral pandemic, governments can quite easily cut funding in public health and vaccine manufacture: invisible services are the first to go. A major concern post-pandemic is that in order to restore the public purse, these items will be high on the list to go, again.

Persuasion can be problematic when the scientists lack the training to speak to the public and the public lack scientific literacy. It is also worth noting that most of the time the 'scientists' on the telly will be representing their own opinion. There isn't one coordinated Corps Scientifique – there are differences in opinions, as becomes immediately apparent if you sit in on a faculty meeting, with heads of department acting as adept cat herders.

Ultimately, because the news is unwelcome, it can be much easier to blame the messenger than the straw-haired, affable clown whose dithering incompetence accentuated the mess. However, I think things are getting better. Lessons learned from the GMO debate led to better communications training for scientists, with organisations coordinating responses (the Science Media Centre in London does a great job of this). And while I've never seen so much scrutiny over every single thing said by scientists as this year, it feels like most of the time the message was put across in a clear and simple way.

Of course, rational science-based arguments failed to win everyone over. So, if on reading this you are inspired to convince racist Uncle Alan over the Christmas turkey that vaccinologists aren't injecting tiny microchips into people, then here is a very short guide. It's not enough to come up with a reasoned argument: the simple fact that microchips small enough to fit down a vaccine needle (0.5 mm, about the width of a human hair) don't exist doesn't seem to persuade people set in their tracks. Nor does outsmarting them by pointing out just how much of their privacy they sacrificed researching mad ideas on their 5G phone. This will simply alienate them. Though you could take the approach used by Winston Peters (deputy prime minister of

New Zealand at the time): when confronted by a COVID-19 denier he simply shut them down by saying: 'Sorry sunshine, wrong place.'[8]

The simplest step is not to engage – do not feed the trolls. Engaging can draw attention to the false ideas and people are equally likely to remember the falsehood as your carefully crafted argument. It's better just to block the bad players and state the truth without reference to the lies. Of course, blocking doesn't work in all circumstances. There will be interactions where you might need to persuade people in real life. Which brings us back to Alan, who by now is probably onto his third sherry, halfway through the Quality Street (taking all the good ones, no doubt) and well into his third conspiracy of the afternoon. To persuade him, you need to link your arguments to his world view, or to public figures he respects, so try reminding him of Margaret Thatcher's chemistry training. Also, use narratives to link the arguments to stories, explain that not getting a cholera vaccine could lead to him passing eighty litres of rice-water and finally put the onus back on him to produce conclusive evidence to support his ideas or at least find out what evidence would dissuade him from his view. If all else fails, buy him a copy of this book, because then at least I get something out of it!

At the end of the day, as a scientist I am inclined to trust science. Working with infectious disease doctors and vaccinologists, I can see very clearly that there is no conspiracy. I have been told directly that my colleagues and I are either directly evil or tools of a broader power. This is just not true. We want to cure diseases and prevent future infections, simple as that.

And with that message we near the end of the story, but before we finish I am going to speculate wildly about what the future might hold in the prevention and control of infectious disease. The following chapter will cover some of my predictions about our fight against pathogens. But before you read it, I want to strongly caveat that these are basically just informed guesses. Firstly, because my track record of predicting breakthrough technologies is very poor. If there is a soon to be obsolete technology, that will be the one I would invest in (minidisc player anyone?). And secondly, because 'the next big thing' will most likely come from a completely unexpected and neglected corner of the research world.

CHAPTER 15

The Future

Timeline: 8 December 2020. Epsom. First dose of licensed Pfizer/BioNTech vaccine given to Margaret Keenan, second dose given to William Shakespeare. Global COVID-19 cases 65,872,391; deaths 1,523,656.

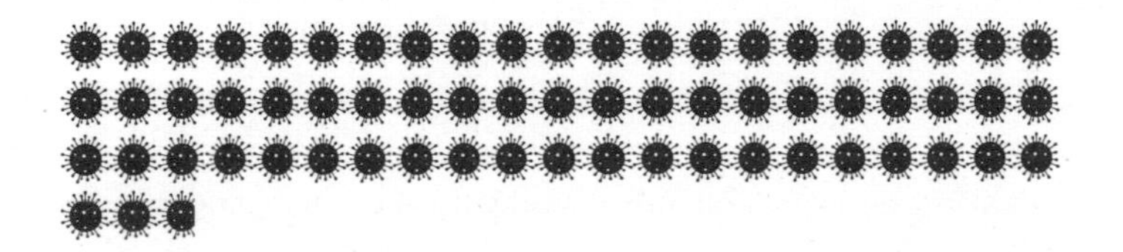

'Even the darkest night will end and the sun will rise.'
Victor Hugo

WHILE HE HAD to deal with some unusual parts of the job – space aliens, teleporting, Captain Kirk's too-tight onesie – Dr McCoy had it easy on one level: the tricorder. With this single gizmo old Bones could diagnose any infection, injury or disease, even in aliens that looked suspiciously like people dressed in rubber costumes. Such is the impact of the tricorder that the XPRIZE organisation, who fund innovative research, started a competition to develop platforms that could diagnose multiple diseases. In this chapter I will look at possible innovations that might play a part in controlling future outbreaks.

HARDER, BETTER, FASTER, STRONGER

Diagnosis will continue to change dramatically. New technologies will certainly play a role in accelerating diagnostics in the future. Even within 2020 we saw a step change increase in the speed and availability of diagnostic tests. In the future, faster, more accurate diagnosis will have a huge impact on how we deal with pandemics. If you can be screened as negative within fifteen minutes in a queue for an event, then live sport and music can start again. Given Britons queued for thirty minutes to get loo roll in March 2020, I can see them queueing for fifteen minutes to watch a cricket match in 2021.

Many of the current approaches look for evidence of the pathogen itself. But we will also get better at diagnosing pathogens by means of the individual's response to them. The body controls infection with a whole programme of responses (see Chapter 5) and these can potentially be interrogated to indicate that a person is infected. At a molecular level, this programmed response to infection can be measured in changes in the person's RNA (the messenger molecule that tells the cells what proteins to make). We can read all the messenger RNA in a sample (the transcriptome), which gives us a snapshot of what the cells are doing at any time point. The transcriptome can be used diagnostically. For example, Dr Jethro Herberg and Dr Myrsini Kaforou, working at Imperial College with Professor Mike Levin, observed a difference in the blood transcriptome in children after a viral infection compared to a bacterial one.[1] This difference in signature can be used to inform treatment choices; for example, bacterial infections respond to antibiotics and viral ones don't. Identifying the genes that differentiate between viral and bacterial infections requires computational analysis and notably Dr Kaforou's background is chemical engineering, not biological science – computers will be central to the next generation of diagnostics.

Wearable technology (Fitbits and the like) may well change how we track infections. By sensing changes from baseline levels of activity, sleep patterns or heart rate, the wearables might be able to detect infections. Fitbit reported that they could predict COVID-19 infection with a 77% accuracy based on increased respiration and heart rate.[2] With more biometrics recorded, diagnostic accuracy will increase.

Another way of tracking disease spread is through interrogating people's search history. Teams at Google tried to use anonymised but geographically based search data to investigate the spread of influenza infections. They had some success, but it still needs some refinement. There are limitations as to what anonymised 'Big Data' can do. The old problem of correlation does not imply causation has reared its ugly head; search terms relating to high-school basketball spike at the same time as influenza because the flu season peaks when the school basketball season climaxes – dunking does not cause flu. Research is ongoing, however, and some search terms correlated with the frequency of COVID-19.[3]

Side by side with Big Data, the digital health revolution will be driven by artificial intelligence (AI). I cannot begin to understand how this works, but computers powerful enough to beat Garry Kasparov at chess by spotting patterns will probably be able to spot patterns in infectious disease. AI could play a role in several spheres relating to diagnosis: detecting outbreaks of infection, forecasting future spread and contact tracing. But there are almost certainly other unthought-of roles for AI in the control of infectious diseases: antigen and drug design; identifying sources of misinformation about diseases; predicting risk susceptibility; writing my grants for me. Combined with the data coming from the microchips Bill Gates has injected into everyone with the COVID-19 vaccine, AI will be unstoppable.*

The final component of the digital revolution is robotics. To date, robots in science labs have been disappointing. They can barely move small volumes of fluid up and down or take the lids off tubes, let alone dance. I am vaguely opposed to intelligent robots in the lab as that would clearly put my job at risk. But there is likely to be some level of automation, particularly on the diagnostics front. The Lighthouse Lab at Imperial College established by Professors Graham Taylor and Paul Freemont had real-life robots, though it failed to give them comedy names: missed opportunities included PCR-McPCRface, Queen ELISAbeth (after an assay named ELISA) and GLaDOS (the cake is a lie).

But one caveat with tech is that if it is too good to be true, it probably is. Before you go throwing your money into diagnostics companies that promise to change the world, do some deeper research. The company Theranos promised breakthrough blood tests that could diagnose a wide range of problems, but it ultimately turned out to be a largely fraudulent venture using existing platforms to produce the data.[4]

* Joking, obviously.

THE FAULT IN OUR GENES

Another way in which the control and prevention of infectious disease will improve in the future is in the use of genetics: gene sequencing and gene engineering. The cost of sequencing all of the genes in an individual person has dramatically fallen over time. This can allow the identification and prediction of susceptibility to infection. It will only fall further in the years to come. Genetic mapping can also be linked with family trees to understand the roots of disease. In Iceland, which admittedly has a very small population and a very well-known genealogy, the genetics of every single person is known. This enables incredible understanding about why some people get sicker than others.

Knowing your genotype could be used to tailor individual risk profiles to specific infections and then to tailor individual vaccinations. If I know I am unlikely to be infected with chickenpox but am at severe risk of dengue fever I could choose accordingly. At a population level this could be used to prioritise vaccines for those who need them most.

Mapping people's genes is just the first step. The next massive breakthrough will be the ability to fix any problems using gene therapy. In the last thirty years gene therapy has moved from science fiction to science fact and it will become more common as the price decreases and the delivery methods are improved. The principle is simple. Some diseases are caused by a faulty copy of a single gene, which leads to an increased susceptibility to infection. If this faulty gene can be replaced by a working copy, then the disease can be cured.[5] An example of a single-gene disease associated with increased susceptibility to infection is cystic fibrosis (CF), which is caused by a mutation in the CFTR gene (described in Chapter 3).

One example of successful gene therapy is to fix immunodeficiency. Genetic immunodeficiency comes in a wide range of guises (due to the complexity of the immune system, there are lots of genes that control it). Immunodeficiency can be broad, for example severe combined immunodeficiency (or SCID), which occurs in about 1 in 100,000 births. SCID most commonly occurs when the B and T cells fail to develop properly. It is sometimes called bubble baby disease after David Vetter, who lived out his short life in the 1970s behind a protective bubble to prevent infections getting in.

One gene associated with SCID is adenosine deaminase (ADA). The ADA gene encodes an enzyme that breaks down one of the building blocks of DNA: deoxy-adenosine. Too much deoxy-adenosine in the cell prevents DNA production and because T cells replicate more rapidly than other cells they are more susceptible to conditions that affect their ability to make DNA. Patients with ADA have fewer T cells, predisposing them to infection. GSK have recently licensed a therapy for ADA deficiency called Strimvelis. Gene therapy requires several steps. First, cells with the potential to develop into T cells (called progenitor stem cells) are collected from the bone marrow of the patient. The collected cells are treated with a modified virus which inserts a working copy of the ADA gene and they are then returned to the patient, where they replace the broken T cells.

To date, gene therapy has only targeted the most common conditions, which makes sense given the frequency of genetic disease and the expense of bringing a single drug to market. Diseases in any one gene are relatively rare.

While approximately 1% of all children born have a single-gene disease of any kind, there are thirty thousand human genes, so the number of children born with defects in the same specific gene is much smaller. Somewhere in the future it may well be possible to get personalised genetic medicine by tweaking the genome to the fittest possible. This is said with no consideration of what the ethics would look like: complicated, I guess!

The tools for gene engineering will also increase in sophistication. CRISPR gene engineering technology, which allows targeted gene editing, has been used in so many different systems that it seems inevitable it will be used in humans in the near future. This was attempted, illegally, when Dr He Jiankui tried to reduce susceptibility to HIV infection by editing the CCR5 gene and it will no doubt be done again. Whether this will be specifically for diseases associated with infection or more general conditions is not clear.

One CRISPR-based genetic engineering approach to prevent infectious disease is called a gene drive. Rather than fixing broken genes in people, the gene drive targets the vectors that can spread the pathogens. Gene drives aim to introduce a mutation that stops *Anopheles* mosquitoes (the malaria-carrying ones) from breeding. Through a piece of genetic wizardry, the gene drive mutations are engineered so that they can spread rapidly through the whole population, leading to a crash in numbers. This approach has been shown to work under lab conditions but the next step of a field trial is still some way off, because there are legitimate concerns about the knock-on effects of this technology – once the genie is out of the bottle, it may be hard to put it back in.[6]

Other innovative approaches for mosquito management are also being developed. *Aedes aegypti*, the mosquito vector that spreads dengue fever, is susceptible to infection with a bacteria called *Wolbachia*. Bacterially infected mosquitoes do not spread viral infections to people. The Global Malaria Programme team have deliberately infected mosquitoes with this bacteria and released them into the wild: in 2020 they reported a 77% reduction in the spread of dengue in the city of Yogyakarta, on the island of Java – a staggering reduction. The aim is now to increase the roll-out of this intervention.

THE CURE

In addition to better diagnostics and genetic engineering, there will almost certainly be breakthroughs in the future in how we manipulate our immune systems to prevent and treat infections. One approach that has roots in the work of Ehrlich and Glenny is therapeutic antibodies. Passive immunisation – the transfer of convalescent serum (blood from a recovered patient to an infected one) – returned during COVID-19 with mixed results. There is an alternative to the relatively crude approach of taking a cocktail of antibodies from one person and sticking them into someone else and that is to isolate *the* single antibody which prevents the infection. César Milstein and Georges Köhler made a major breakthrough in this process while working at the MRC Laboratory of Molecular Biology in Cambridge in 1975.

Köhler and Milstein wanted to mass-produce antibodies from B cells. They knew that each individual B cell produced a unique antibody, but when they took a sample of blood they got a cocktail of B cells and therefore a cocktail of antibodies; no better than transferring convalescent plasma. Köhler and Milstein solved this problem by fusing individual B cells with cancer cells to produce immortal cell lines. Crucially, each cell line produced only one specific antibody, which they called monoclonal antibodies. Monoclonal antibodies (mAbs) have changed *everything* in biology, rightfully earning Köhler and Milstein the Nobel Prize in 1984, just nine years later. I would argue that they are as *important* a breakthrough as either antibiotics or the microscope. Monoclonal antibodies find a use in a huge range of diagnostic tests, experimental settings and, most relevant to this chapter, therapies. There are at least two ways in which monoclonal antibodies will be used in the control of infections: prevention and immunotherapy.

PREVENTION INVENTIONS

Antibodies can be used to directly target and kill pathogens. Antibodies are a very precise form of medicine: they will bind to the pathogen and nothing else. This means they can be used with a high degree of specificity. To date there are three antibodies licensed for use against infections in humans: palivizumab against RSV and raxibacumab and obiltoxaximab against anthrax. Of the licensed antibodies, palivizumab is most widely used because RSV is a much more common infection.

Monoclonal antibodies can potentially combat pandemic infections. The 2014–16 Ebola outbreak saw the emergency use of ZMapp, a cocktail of anti-Ebola virus antibodies, in infected individuals. The antibody was pharmed in transgenic tobacco plants (by a subsidiary of British American Tobacco). However, the approach was highly experimental, stocks ran out in the first wave of the outbreak and then the number of cases declined to such a point that it became impossible to perform clinical trials to properly demonstrate efficacy.

One of the experimental treatments that 'the Donald' reportedly received after recklessly catching COVID-19 was an antibody cocktail produced by the biotech company Regeneron. Regeneron use an interesting technology to isolate their antibodies. They have a mouse that has been genetically engineered to make human antibodies. This means the antibodies produced are more likely to be tolerated by patients (getting around some of the problems Ehrlich had with horse serum in the 1920s). Their treatment received emergency use authorisation in November 2020 and Regeneron announced promising data in December 2020, but as with many of the other novel therapies in the COVID-19 era it was not licensed at the time of writing.

Other companies also produced antibodies to treat COVID-19, including Eli Lilly, which developed the completely unpronounceable bamlanivimab. This follows a series of other monoclonal antibodies with bonkers names, including tocilizumab (anti-interleukin-6), infliximab (anti-tumour necrosis factor) and palivizumab (anti-RSV). I'd always thought the names were created by some kind of random letter generator or just by asking a toddler to hit random keys. But it turns out there is an international naming committee for drug names. The WHO coordinates this with a mandate to make sure that the drug you are taking is what you think it is, so you ingest paracetamol not paraquat. A WHO sub-committee set a standard for naming monoclonals: the first part is described as a 'fantasy' name, the second part relates to function and the ending is mab.

It's a bit like those memes – what's your monoclonal name: first part is first pet's name; second part is disease area. Mine would be cat-vi-mab (yes, our cat was called cat; we couldn't decide on a name so we probably needed a committee). So with the anti-COVID-19 antibody, Eli Lilly chose 'bamlani'; 'vi' indicates it is antiviral and 'mab' that it is an antibody. I find it pleasing that people have spent time working this all out; and doubly so that it wasn't me who had to sit on the subcommittee while it was discussed, no doubt at length.

The demonstration that monoclonal antibodies can potentially prevent disease during COVID-19 is a huge boost to the field. Possible uses include controlling chronic infections such as HIV or preventing antibiotic-resistant bacteria in hospitalised patients. Another thing that will affect the deployment of antibodies is a reduction in cost. The anti-RSV antibody palivizumab is too expensive to be used on all children, so it remains reserved for very high-risk babies. Bringing the price of antibodies down will have a huge impact on their use, filling the gap for when vaccines are not yet available. One potentially cheaper and longer-lasting antibody therapy for RSV is called nirsevimab (or MEDI8897). It is being developed by Astra-Zeneca and Sanofi and is hopefully coming to a maternity ward near you soon.

IMMUNOTHERAPY

Antibodies can also be used to manipulate the immune response. Your immune response is important in stopping you getting infected, but too much immunity can be a problem too. This is most commonly seen in autoimmune conditions such as arthritis. One of the hallmarks of autoimmunity is an excess of cytokines in the blood.

A breakthrough discovery about the role of cytokines in arthritis was made at Charing Cross Hospital in London.* Marc Feldmann, born to a Jewish family in Lvov, Ukraine during the Second World War and Ravinder Maini, an Indian-born British rheumatologist, led the project, reiterating the international make-up of science. Feldmann and Maini identified much higher levels of a cytokine called tumour necrosis factor (TNF) in people with arthritis. They then developed a monoclonal antibody that blocked the cytokine from causing inflammation, with remarkable results; it led to the regression of arthritis in many patients.

So how is this relevant to pathogens? The immune system drives many of the symptoms associated with infection, for example a temperature. Sometimes the immune response is too strong and in the process of clearing the pathogen it overshoots and starts attacking the body too. This is described as infection-induced immunopathology and it is potentially a target for treatment. The same cytokines that can cause autoimmunity may also drive disease after infection. By damping them down, it may be possible to reduce the disease associated with infection.

I think there is also space for these kinds of immunomodulatory treatments in the control of infectious disease. One of the first drugs to be shown to have any effect on severe COVID-19 was dexamethasone, a corticosteroid. A class of hormones, corticosteroids were first isolated

* Charing Cross Hospital is confusingly located in Hammersmith, London and is not to be confused with Hammersmith Hospital, which is in White City, presumably to screw with health tourists.

from the adrenal glands in the 1940s. The early natural compounds were used to treat rheumatoid arthritis and were then replaced with synthetic mimics, because they were easier and cheaper to produce. Dexamethasone was one of the earliest synthetic corticosteroids. First licensed in the 1950s, it is used in a wide range of diseases, particularly autoimmune conditions in which it dampens the inflammation. This same action appears beneficial in COVID-19 and thanks to large, multi-centre clinical trials (RECOVERY run by Oxford University and SOLIDARITY run by the WHO) and the patients who volunteered to be part of them it was possible to see what did and didn't work.

Dexamethasone is a bit of a shotgun approach, hitting multiple targets. A more targeted approach, like the use of anti-TNF for arthritis, might be more beneficial. For example, the cytokine interleukin-6 (IL-6) has been linked to severe COVID-19 disease. Like TNF, IL-6 is pro-inflammatory, increasing the activation of other immune cells. Two large studies showed that using antibodies to block IL-6 could reduce severe disease, opening this path for future interventions. This builds on many years of pre-clinical studies that had worked well in animal models but never quite translated into benefit in people. One consideration had always been timing. In the lab we know exactly when the infection starts so we can time the treatment more accurately. People have the annoying tendency to react differently to infections, but with better diagnostics of both the pathogen and the response to it we might be able to pin down the exact time of infection and therefore tailor the interventions.

However, one consideration for suppressing the immune response is that immunity is quite important in stopping infections. Following TNF blockade some arthritis patients came down with tuberculosis. These patients most likely had latent TB held under control by their immune systems, specifically by cells producing TNF – when this was disrupted the bacteria awoke from latency. Future interventions looking at altering the immune response will need to be mindful of these unexpected effects.

FIGHT THE POWER

As well as synthetic compounds that mimic human functions or modulate the immune response, microorganisms themselves will most likely be called up to fight infections. The use of microorganisms as therapy has a long history. The simplest use is to kill the pathogen directly. Research into the use of phages to kill bacteria has been reinvigorated, which would no doubt tickle Félix d'Hérelle's fabulous moustaches.

Microorganisms can also be co-opted as a stimulant to the immune system. William Coley, an American cancer doctor working at the turn of the twentieth century, noticed that when one of his cancer patients went into remission after being infected with *Streptococcus pyogenes*. He tried to replicate this finding by deliberately infecting another patient's tumour with *S. pyogenes*; remarkably, this worked and the patient also went into remission. Coley used his bacterially derived toxins on other patients with variable outcomes, accidentally killing two out of ten patients (not a great success rate). Regardless of the success rate, Coley's approach has influenced thinking about utilising the immune response to treat cancer. A more targeted approach in cancer immunotherapy through the use of monoclonal antibodies has been transformative in the last five years.[7]

Another area that will almost certainly grow is the manipulation of the microbiome in the context of disease. A huge amount of time and energy has been spent trying to understand the microbiome. I think we are at the stage where we can confidently say it has a role in human health, but not yet at the stage where we can specifically alter it in a way that has a predictable outcome. As microbiologists get better at isolating the different bacteria in the cocktail of bugs, they will be able to make more targeted products for the treatment of a range of conditions, potentially including preventing infections.

TRIGGER INSIDE

Infectious pathogens can have a longer legacy than the initial disease. The future will no doubt see leaps in our understanding of the links between communicable and non-communicable disease.

It has long been known that infections can trigger cancer. Peyton Rous first demonstrated this in chickens, showing that cancer could be transmitted between chickens by injecting ultra-filtered material from one bird to another. Tony Epstein and Yvonne Barr followed up on Rous's work, isolating the first human virus associated with cancer from individuals with Burkitt's lymphoma. The virus they discovered (Epstein-Barr virus) is mostly associated with glandular fever, the disease that affects teenagers because it is transmitted in saliva (and by inference heavy petting). Since then several viruses have been associated with cancer, including HTLV, HBV and HCV. One of the most well-known links between infection and cancer is human papillomavirus (HPV). In 1974, Harald zur Hausen hypothesised that HPV caused the genital warts that led to cervical cancer, for which he shared the Nobel Prize with Françoise Barré-Sinoussi and Luc Montagnier (for their discovery of HIV). Nubia Muñoz, a Colombian epidemiologist, confirmed Hausen's hypothesis. She identified the strains most closely associated with cancer (HPV16 and HPV18), which led to the development of the incredibly effective vaccine.

But looking forwards, other long-term consequences of infectious diseases may be identified and potentially prevented. In particular, autoimmune conditions may have an as yet unidentified infectious trigger. For example, a condition called rheumatic fever follows an infection with group A *Streptococcus* (*S. pyogenes*) in some people. The proteins on the surface of the bacteria mimic human proteins and so an immune response against the bacteria may end up attacking the human tissue as well. It is not uncommon for pathogens to try and look as similar as possible to human cells, in order to escape the all-seeing eye of the immune response. In the process of fighting off these pathogens the body can accidentally target itself. One of the challenges going forward will be to identify microbiological triggers and determine whether they play a role in a wider range of conditions, for example type 2 diabetes, without falling into the correlation/causation trap.

Infection may also affect our mental health. Edward Bullmore, a psychiatrist at Cambridge, has proposed that depression might be linked to inflammation.[8] As inflammation is associated with infection, it is not a huge step to suggest that infection can trigger depression. During the COVID-19 pandemic, a group in the United States published data suggesting that survivors had an increased risk of psychiatric events.[9] There may be an evolutionary component to this, with infection leading to behaviours that separate survivors from the herd to prevent spread, but this is all hyper-speculative and will need a lot more in-depth study to finalise it.

#VACCINESWORK

Of course, the best way to manipulate the immune system to prevent infections is vaccines. Returning to the vaccine timeline in the last fifteen years, five new vaccines have been licensed. This is a huge achievement, especially as prior to 2005 vaccines had been seen as a research field in decline. The renaissance in vaccines in the last fifteen years also makes me optimistic for the future of antibiotics. Without doubt, vaccines will continue to be part of the infectious disease tool kit for the next hundred years and I fully anticipate there will be further breakthroughs in the following areas:

- Vaccine Development Time. We will see a further acceleration in speed from lab bench to patient bedside. This will build on the amazing speed of development in 2020 – just to reiterate, this was astonishingly fast. Pfizer/BioNTech released the first evidence that a vaccine could prevent COVID-19 in November, less than ten months after they had started their programme. The first doses were going into the public a month later. Professor Robin Shattock (Imperial College) is optimistic (always) about how RNA vaccines will shape pandemics of the future: '2020 is just the beginning of unlocking this transformative technology.'
- Global Access. Vaccines need to be accessible to all people around the world, particularly people in low-income

countries. Advances in two broad areas are needed to improve global vaccine equity: better stability and devolved manufacture. In terms of stability, the problem is that most vaccines are heat-sensitive – when left at room temperature for more than a short time, they stop working. Making vaccines that can be stored at room temperature will mean more of the globe can access the vaccines they need. With regards to manufacturing, alternative approaches will be developed which could distribute manufacturing more globally. One attractive approach is to use genetically engineered plants to grow vaccines. I spent my early twenties working on this idea,[10] but then changed fields, so it was very satisfying to see the area blossom recently with a large clinical study demonstrating the safety and efficacy of a flu vaccine grown in tobacco plants;[11] and, of course, a chance to roll out some evergreen puns.

- Globally Distributed Manufacture. For a number of reasons the diseases that infect people in low-income countries receive less attention; one of which is simple economics, particularly the recovery of development costs. A huge change in the last forty years is the global redistribution of vaccine manufacturing. Manufacturing at scale can reduce the cost of the vaccine, making targeted vaccines for low- and middle-income countries possible. The company that manufactures the most doses of vaccines in the world is the Serum Institute of India. Working with the Gates Foundation and SynCo Bio Partners (a Dutch biotech), the Serum Institute of India manufactured MenAfriVac in response to a surge of meningococcal A meningitis that swept across sub-Saharan Africa, killing 25,000 children in 1996–7. The vaccine was rolled out in 2010 at less than US$0.50 per dose. It has had an amazing impact. MenAfriVac reduced meningitis incidence in Chad by 94% and by 2013 only four cases of MenA meningitis were detected across twenty-six countries.

A positive outcome from the COVID-19 pandemic has been the re-establishment of national level vaccine manufacturing capacity, which may well build on novel vaccine approaches like RNA or viral vectors. These platforms may be more suitable to small, local-scale manufacture. It might be possible to fit everything into a shipping container as part of a rapid response kit to be deployed in parallel with Ian Goodfellow's 'hydroponic' virus sequencing tents.

- Better Understanding. There are still so many questions about how the immune response to vaccines really works; for example, adjuvants (the substances added to vaccines to boost their immunity) are still a bit of a black box. Underpinning better vaccines will be better models of infection going from single cells to models of whole human organs, integrating understanding from natural infections, animal models and deliberate human infection.
- The Big Three. Vaccines for HIV, TB and malaria are needed and some progress is being made in this respect. There is definitely a new TB vaccine on the horizon, which promises to be more effective than BCG. The RTS-S vaccine indicates malaria is a solvable problem. HIV still looks extremely challenging, but there's no reason to think the combination of innovation and investment into this area won't ultimately pay off. After all, malaria was discovered one hundred years before the

vaccine was developed and at forty HIV is a relative youngster.

- Eradication. With a concerted effort, I think we will see the eradication of another virus by 2030 – polio. Wild type polio has been eradicated from Africa and is endemic only in Pakistan and Afghanistan. This will be accelerated by the new, more stable oral polio vaccine. Looking further ahead, there is no reason to think other viruses can't be eradicated: measles has all the characteristics of an eradicable disease, so let's eradicate it.

The future for vaccines is really positive. COVID-19 demonstrated the impact vaccines have in restoring life to normal and for every crazy person who believed Bill Gates wants to individually track us using imaginary technology there were a hundred people who just wanted to take a COVID-19 vaccine so they could return to the pub and see granny.

So, what does the future hold? As with most Trekkie technology (except maybe Spandex jumpsuits and communicators) I think we are a way off tricorders. But in the next decade I predict there will be revolutions in diagnostics, vaccines and gene therapy. Progress can be extremely rapid if the funding and political will is there. One thing is for sure; if the progress is as fast in the next one hundred years as it has been in the last hundred then 2121 is looking pretty fine.

EPILOGUE

Where do we go from here?

Timeline: 20 December 2020. Epsom. Inevitable tightening of UK restrictions begins (FFS). Christmas cancelled. Global COVID-19 cases 76,703,623; deaths 1,693,585; vaccine doses 2,760,000.*

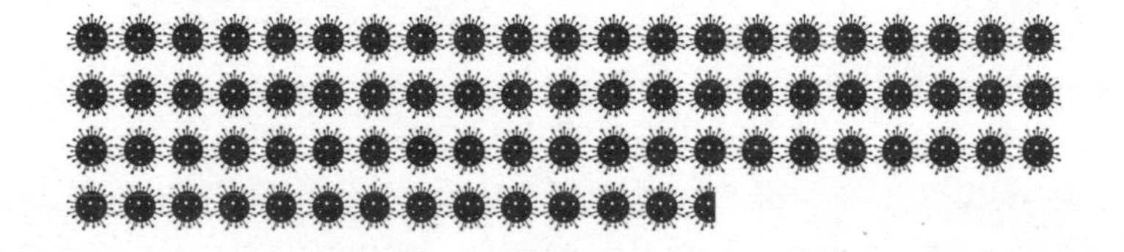

'Messieurs, c'est les microbes qui auront le dernier mot.'

Louis Pasteur

THIS BOOK HAS been shaped by the COVID-19 pandemic, which made it clear to everyone that infectious diseases possess the power to bring the whole world to a screaming halt. Pathogens, if uncontrolled, can have an enormous impact on our world far beyond health, affecting the

* Each represents one million COVID-19 vaccine doses.

economy, politics and even the environment. As I finished this final chapter in the dark on the shortest day of 2020, with the virus surging around us and the emergence of new variants dominating the news, it was clear we had all been on a journey. Even those (idiots) who didn't believe that a virus caused COVID-19 had their lives upended by it.

Knowing lots of random facts about infectious diseases didn't prepare me much better than the tin-foil hatters. I was marginally ahead of the toilet-paper purchasing curve and did somewhat better in picking places to go for a summer holiday in 2020. I was also able, eventually, to persuade my parents that it was going to be quite serious and they needed to take care – though, on reflection, telling my mum at a school concert that everyone in the audience could be dead by the end of the year was a bit of an overstatement. For all that knowledge, I was not prepared for the disruption of lockdown at the beginning of the year.

I WASN'T THE ONLY ONE

To try and put 2020 into perspective, I spoke to scientists at the heart of the maelstrom. Professor Wendy Barclay, an influenza expert and member of SAGE, the UK government's scientific advisory group, said:

> We have just experienced historic times. I have spent my entire scientific career telling the public, politicians and funders that a respiratory virus pandemic would cause massive disruption. But in all honesty, I never actually understood how it would *feel*. This pandemic has disrupted and challenged everyday life in a way that no other event in my own lifetime has touched me. I now see that what affects

> us as individuals are not the events themselves, not what *happened* but rather how their impact is *felt*.

Professor Steven Riley, a member of SPI-M (Scientific Pandemic Influenza Group on Modelling), told me how 2020 for him brought back echoes of 2003 (the first SARS outbreak):

> For three sleepless weeks in March 2003, we imagined what a coronavirus pandemic could look like. Luckily, that didn't happen then. Unfortunately, in 2020, a lot of what we worried about in 2003 came true. In those early days in 2020, I think there was a small group of niche specialists who had looked at outbreaks and pandemics since 2003 in a particular way and were able to give useful insights to directly help the response and to prioritise analyses and data collection. On a scientific level it was exciting and extremely interesting, but also incredibly stressful.

Professor Riley wasn't the only scientist to find 2020 exciting. Robin Shattock told me that: 'I've worked on vaccines for twenty years and never had a year that has been so intense and so exciting.'

BE PREPARED

Hopefully, no one will forget the dramatic impact that pathogens can wreak upon the world. We are not out of the woods. I cannot promise the global infection event we have just experienced will be the last. As the COVID-19 outbreak shows, we are susceptible to outbreaks of pathogens and their hugely disruptive impact on the world. With demographic shifts and climate change we are likely to face other pandemics. There's another storm coming and another one after that, though hopefully not to the scale of COVID-19, at least not in my working lifetime (my kids can deal with the next one). So, what are the lessons I have learned, that my children can apply?

We must be ready for the next outbreak. Some of that requires guessing what it will be. Two prominent threats exist: antibiotic-resistant bacteria and zoonotic viruses. Both threats underline how closely we humans are tied to the world around us and how, for all our innovations in health, our actions in other spheres put us at risk. Some of these actions are inherently avoidable: burning through our front-line antibiotics to make chickens grow faster; spraying tonnes of the only antifungal drug that actually works across fields of tulip bulbs; growing millions of mink for fur only to find that not only can they be infected with SARS-CoV-2 but they return the compliment in spades, inducing mutations in the virus that threaten all the hard-won progress we have made in developing vaccines and drugs.

All of which pales into insignificance alongside the huge impact global heating will have on the spread of infectious disease. We are seeing this already – the normally benign fungi *Candida auris* is beginning to adapt to the warmer climate and as a side effect is adapting to our body temperatures, removing a barrier to infection; warmer climates mean a bigger range for mosquitoes and all the 'orrible bugs they carry; and changing environments also put humanity in closer contact with other species – particularly bats, with their cocktail of viral pathogens. All of which is deeply depressing.

However, let's emerge into the post-COVID-19 dawn with resolve. There is hope and it comes in the form of preparedness, innovation and research; all of which require investment. In the context of the trillions lost in a single quarter in 2020 and the massive disruption to the global economy, investment in pandemic preparedness is peanuts. The UK spent £5 billion on Trident and we, as sure as shit, never want to use that. Let's spend an equivalent sum on a pandemic plan that we may never use. The investment needs to be made across the piece. We need better hospitals with actual spare capacity, which means hospitals don't need to cancel your cancer operation to save your granny from a respiratory pandemic. We need a working public health system that is centrally coordinated but locally administered. We need spare manufacturing capacity for biologics, or ideally capacity that can be quickly repurposed. This can, with some innovative thinking, be done in a way that the capacity is not gathering dust; gene therapy uses the same technology as vaccines and cancer immunotherapy uses the same manufacturing plant as antiviral antibodies.

Unfortunately, in 2020 we were caught short. Ten years of austerity budgets had seen to that. The Tory governments basically cancelled the health insurance payments; which *may* have made economic sense in the short term, but gambling against future risk is not necessarily the job of responsible government. In *The Black Swan* Nassim Taleb describes the surprising frequency of rare, extreme events and their disproportionate impact. The book takes its title from the fact that black swans exist, even if you have only ever seen white swans; absence of evidence is not the same as evidence of absence.[1] From this comes the need to build resilient systems, or at least not just bury one's head in the sand and hope that a once in a lifetime event doesn't happen on your watch.

Within this resilient system, I would like to make a case for investment in science. But who should foot the bill? The cost of resilience needs to be shared between the private sector, the extremely rich and the government. With the private sector the case is pretty clear (to me): pay a small amount now or pay big later. In my simplistic economics model a company would be better off paying a smaller percentage each year rather than face total collapse. A proportion of the companies that collapsed in 2020 could have survived had the pandemic been better controlled. It is part of the social contract: the failure of public health in 2020 can trace its roots to the banking crisis of 2007, so it doesn't seem like too big a step to ask the financial institutions that were bailed out in 2007 to contribute now and in the future. But I suspect I am pissing in the wind: it's not as if the airlines are prepared to pay a levy on global heating, even though it will eventually break them.

To the extremely rich, my question would be: what else are you going to do with the cash? You can't take it with you and there is only so much stuff you actually need. Look to Dolly Parton, who funded the initial research that underpinned the Moderna COVID-19 vaccine. Or Henry Wellcome, whose foundation has done so much in infection. And, of course, to Bill Gates, who unlike some of the other uber-rich has given much of his money away and has pledged to donate it all before he dies. He has made a remarkable, positive impact on the world. To date the Bill and Melinda Gates Foundation has spent $53.8 billion, critically using it to coordinate and leverage government funding. The campaigns to eradicate polio and malaria owe a huge amount to funding from Gates, as does the remarkable MenAfriVac campaign. The Gates Foundation were also central in the creation of GAVI, which has provided 440 million immunisations and averted six million deaths.

Presuming someone has listened to me, how do we convert the investment into progress? While all science involves repeated testing of ideas, we artificially split the world of scientific effort into two very broad areas: basic science – pure research, learning about stuff for the sake of learning; and translational science – testing things (drugs, chemicals, devices, bridges, computers) to improve the quality of human existence. To those with a commercial mindset, the latter, translational, approach has the greater value. You put money in and you get better stuff out. This can be seen with the vaccines, where investment can pay dividends, not just in health but also financially: BioNTech floated in October 2019 at $11 a share and they are now worth $110 a share – a tenfold increase in a year!

But translational, product-driven research is not the be-all and end-all of science. We also need pure research: the ideas about how to make stuff better come out of pure research. Lots of modern engineering depends on us understanding how gravity works, but Newton's aim wasn't to put rockets on the moon. While the results are not immediately tangible, basic science underpins technologies that are the foundations of billion-dollar industries, for example cancer immunotherapy, antibodies, lasers, the internet, GPS, etc. Funding translational science at the expense of basic science may pay off in the short term, but it damages advances in the long term. The innovation breakthrough of tomorrow is currently festering in an obscure underfunded lab out there. CRISPR, the magic gene editing technology, came out of some noodling with the immune systems of bacteria.

This is where I think government funding should come in. In my opinion, universities and institutions are best placed to deliver pure research and companies are best placed to develop the ideas into real things. Of course, there is fluidity, with great basic research in companies and product-driven science in universities. To support this means government money predominantly goes into open-ended research with no obvious product. This division may seem unfair, with governments paying for the basic research and not gaining anything material and the companies not paying anything but gaining from any breakthroughs. But the money will trickle back in tax take, employment, better medicines and other indirect benefits. As seen in 2020, investment in the health sciences has a significant return on the speed of research during a health crisis: the breakthroughs in COVID-19 treatment, vaccines and sequencing were all built on a base of research laid down in the previous decade.

Another benefit of governmental investment in research is in teaching and training – the economy needs people with science backgrounds. A PhD provides the student with very much more than just the ability to move colourless liquids around. It gives them problem solving, teamwork and analytical skills, tenacity, flexibility and independence. The best way to deliver all of this is through science research. Just as no one expects doctors to train without ever seeing a patient, the best way to learn science is by doing science.

This training part is critical, because it is the next generation of scientists who will deliver the next generation of scientific breakthroughs. One of the things I have discovered in writing this book is that often the big man who shouted loudest about the idea isn't really the person who did the work, even though they got the credit. I hope to have reflected some of the diverse cast of experts who chipped away at the mysteries of pathogens and the immune system. Going forwards, we need to support the best people regardless of their background, race or gender and we need to do it globally. Focusing advanced science education on a tiny percentage of the world's population is going to miss some of the best and brightest minds. Better distribution of training is more likely to address local issues with appropriate solutions. Gene therapy at $1 million a pop will not solve the AIDS epidemic in sub-Saharan Africa but education about condoms and cheap generic HAART drugs might.

One final consideration is the difference between cost and value. We need new funding/recovery models for a wide range of health interventions: antibiotics, antiparasitic drugs, the control and elimination of neglected tropical diseases and vaccines. These undoubtedly have huge value in preserving human life, but don't necessarily generate an economic return. Finding ways to decrease the cost of the research and testing, compensate the parties involved and increase the return on the products is vital (though not straightforward). But it can be done – the Oxford/Astra COVID-19 vaccine was made at cost and licensed freely to maximise global manufacturing.

REASONS TO BE CHEERFUL

Preparedness and prevention are the keys to stopping the next pandemic. The great news is that we have made enormous steps in these areas. We only need to look back a hundred years to the 1918 influenza pandemic, which caused

50–100 million deaths, or 5% of the global population. At the time of writing (January 2021) COVID-19 had killed 0.025% of the world's population. Each of those 2.5 million deaths is a tragedy, but it is dramatically less than what could have been. Unchecked, SARS-CoV-2 has a 1% infection fatality rate – if left uncontrolled it could have caused something in the region of 78 million deaths globally. Of course, the pandemic didn't end conveniently as I finished my book; sadly, it will cause more deaths, particularly if the richer nations don't share vaccines with the poorer nations. But it is dramatically less than what could have been.

The progress between the two pandemics has been incredible, with advances in every area: diagnostics, modelling, public health, therapy and vaccines. There has been a huge leap in our understanding. In 1918 the infectious agent was unknown, but in 2020 we had the sequence of the new virus in days. Knowing the causative agent has been critical in controlling the pandemic. SARS-CoV-2 was often referred to as an invisible killer throughout the crisis, but the efforts of the China CDC and the WHO did in fact make it visible. This was then fed into existing and novel tools to track who was and equally who wasn't infected. The knowledge obtained was then fed into the models of how the disease was transmitted, enabling governments to make decisions about how to control spread while trying, but not always succeeding, to walk the tricky line between being too restrictive and too permissive. We will not know the best path until the crisis has blown over – the vital learning from this pandemic is how to prepare for the next one. We must prepare flexibly. The approaches that worked for COVID-19 will not all work for Disease-X. As the Prussian military strategist von Moltke snappily put it – 'No plan of operations extends with any certainty beyond the first contact with the main hostile force'; which is often shortened to 'no plan survives contact with the enemy'. In other words, plans are useful, but flexibility of response is critical.

The high-speed breakthroughs of today are built on a foundation laid years ago. Dexamethasone was developed in the 1950s; the Oxford COVID-19 vaccine was really the final year of a fifteen-year-long project, rather than appearing from nowhere; and BioNTech, the first vaccine company to report a positive vaccine trial for COVID-19, didn't just spring up fully formed – it was founded in 2008.

THE END

'It is time to close the book on infectious diseases, and declare the war against pestilence won.' This famous but sadly entirely made-up quote has been attributed to Dr William Stewart (the American Surgeon General 1965–9). Whoever did coin this phrase captured the considerable inroads humanity has made in controlling infectious disease. People are far more likely to die of non-communicable diseases than infectious ones. While we have made considerable inroads, one thing is absolutely certain: the future holds infectious disease. There is the risk of new pandemics and we mustn't forget the diseases that are endemic globally and the risk that antibiotic resistance poses.

And we certainly don't know everything, not even close. There are still so many questions. How do viruses really transmit – is it by the air or on surfaces or both? Why are some people susceptible and some not? How are the mind and the body linked? Are the patterns of viral mutation predictable? How much of the human genome is vital in controlling infection?

The good news is that with ongoing innovation and research there will be solutions. Coming back to *The Times* article about 1918 influenza, written in 1921, it finished on a scientific note: 'The mystery remains a mystery and a vast field for research is presented to the scientific mind.' Of course, science is often wrong and the ideas that I write here may all turn out to be as true as the four humours. In my study, where I am writing, sits a clock that was given to my great-grandfather as a wedding present in 1901. In the 120 years it has been telling time there has been a revolution in our knowledge about pathogens. I see no reason why this progress shouldn't continue for another 120 years. Through global effort, investment and unity many of the pathogens that infect us today can be eradicated tomorrow. You can be part of this greatest human achievement – get vaccinated, wash your hands and take the right drugs at the right time. Let's be the generation that kicked pathogens into touch. At the end of 2020, a year dominated by pathogens, I remain incredibly upbeat about the progress we have made and hopefully, after reading this, you do too.

POSTSCRIPT

31 January 2021.
Stuck in my house, in Epsom.

Global COVID-19 cases 102,970,000; deaths 2,230,000; vaccine doses 99,280,000.

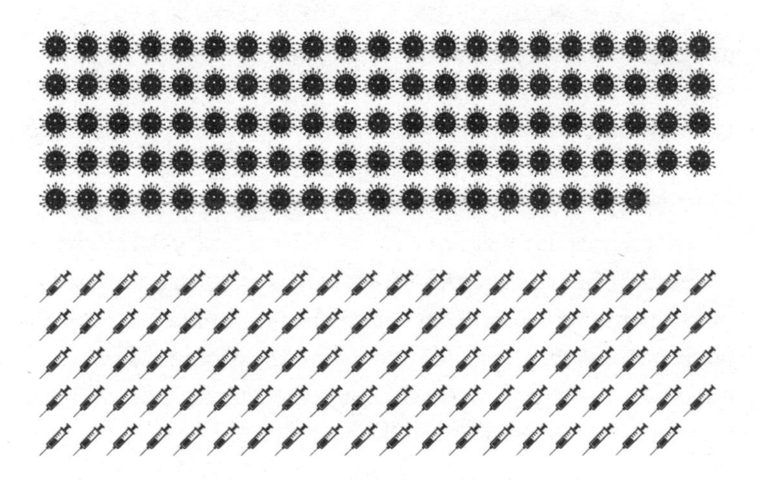

IN THE BEGINNING, I wrote how COVID-19 was inextricably woven through the pages of this book. At the time of writing the introduction I meant it more as a concept; a way of illustrating the science, adding colour and a common reference point.

What I didn't consider was the shared emotional impact of the pandemic and the heightened state of anxiety we all felt through 2020 and into 2021. It ebbed and flowed, but it was always there. Was the virus going to strike me or my loved ones? How would we deal with it? What would it actually feel like to get infected? How would my own immune system cope with it?

This anxiety didn't recede as we learned more about the virus. If anything, knowledge was a burden: I was aware of the odds of severe disease and the risks of long-term effects and knew exactly when I might feel most sick after a positive result.

And the pandemic didn't conveniently go away just because I had finished writing the first draft of my book in late December 2020. In the following weeks the virus got closer and closer to home. First my son had to isolate because his best friend had tested positive, then one of my PhD students went down with it and then our cleaner.

At this point the virus felt close but still conceptual as a risk: a tragedy all too real, but happening to other people, that somehow, for whatever lucky reason, had passed us by.

As 2021 began the crisis intensified, reaching a crescendo of cases in the UK on the weekend of 8 January. It was at some point during this weekend that my wife and I became infected. But it wasn't immediately apparent that we had COVID-19 – we had none of the hallmark symptoms. At the time we both had a sense of smell and we had no fever, while I had a slight cough and a runny nose. To all intents and purposes it felt like a common winter cold. It was only when I had a routine asymptomatic screening test at work that we discovered that I was positive.

Then followed two quite unpleasant weeks and I had only had a mild bout. At the peak of infection it felt like I was at altitude; not breathless but aware of my breath. It was much harder to shout at my kids (to come down to supper) and my singing was curtailed (not a real loss to the world). I would have failed an FEV1 test, the one where you exhale as hard as possible into a toilet roll. I also produced a prodigious amount of luminous green snot. The recovery has been quite slow. I never dropped below 75%, but one month later I am still only at 90%.

It became very clear that anyone who already had problems with breathing would find COVID-19 very challenging indeed. But contrary to expectations I didn't get any of the hallmark features that I had been expecting. I lost my sense of smell for only one evening and I never had a fever. I did get a cough, but I always get a cough, so it wasn't unusual.

The emotional journey was more tiring than the infection itself, especially about seven days in – everyone had told me that the second week was much worse than the first and the first had been pretty unpleasant. It didn't help that a few weeks earlier I had listened to a story on the news about someone's forty-year-old brother who had passed away in the night at home, with little to no warning. It was also very touching that lots of people were checking in on me, but it did nothing to reduce the worry that this was something bad indeed. All of which contributed to some next-level bonkers hallucinogenic dreams; including one dream about being evicted from a pub because a friend had smuggled a skinned rabbit in his trousers.*

Conceptually, one of the odd things was knowing the exact pathogen that was causing me grief – most of the time when I get a cold or a cough I have no idea what caused it. Lockdown and self-containment didn't help, because the

* Interpret that, Freud!

thing you don't need when you are feeling ill is being trapped. Especially when it means having to add home schooling on top of home working, editing a book and not dying. It was exhausting. One small upside was that it made dry January much easier to tolerate.

As I found out in 2008, when my son was hospitalised with RSV, all that knowledge I possessed turned out to be useless at the most fundamental level – getting me and my family better. Yes, I could tell you exactly what was happening down to a molecular level on any given day after infection, but it didn't speed up the recovery. It was all very meta – writing about a pandemic, during a pandemic, while infected with the pandemic. At one point, having woken up from another dream in which I had died, I lay there thinking: 'I have literally written the book about this, so why isn't my subconscious getting the message?'

And then it was over. I cannot stress how fortunate I was to get a mild dose of the disease. I feel a weight very much lifted and I am confident that I am now immune and that if I do get it again it will be milder. I also won't be able to spread it on to my parents, my friends or my colleagues. We had a discussion at work about whether in 2020 it was worse to admit you had given someone an STI or COVID-19. I slept better in the days after recovering than I had done since 2019. There was also a huge relief in not having to be quite so paranoid about every social interaction; no more Lady Macbeth scrubbing of the hands.

As January ended, cases were coming down and vaccines were going up. By 3 January, the number of people vaccinated in the UK equalled the number of confirmed infections and by the 31st this ratio was 2:1 vaccinated to infected. Across the world, the numbers of vaccinated passed those infected in early February 2021, with licensure and global roll-out of multiple vaccines. There are still threats on the horizon – the emergence of new variants that can escape the immune system, vaccine nationalism and political shenanigans.

During the twelve months it has taken me to write and edit this book, the worst infectious disease pandemic in living memory has taken place. It will no doubt leave a lasting legacy on the world, but I have great hope that the very darkest days are over, though this will require global cooperation to ensure all people are protected.

AFTERWORD

Post-postscript (one year later): 31 January 2022. Back to the day job at Imperial College London. Global COVID cases 378,400,000; deaths 5,661,152; vaccine doses 4.84 billion: approximately 25.2 million per day.*

'*The only constant in life is change*'
Heraclitus

So here we are, one year on. The advantage of writing further into the pandemic is that it is easier to reflect on what went well (and what didn't). The COVID vaccines were an undisputed success: when used, they led to considerably fewer people in hospitals, though there was a tragic failure to share these vaccines equitably around the globe, with the vast majority going to richer countries. This led to a disproportionate burden of disease in the countries that could afford it the least. How we distribute the dividends of research more fairly remains a pressing question. A move towards local manufacture of vaccines, so that all countries can produce their own, is one possibility.

* From Our World in Data (https://ourworldindata.org/). NB There wasn't enough space to show as images.

Alternative manufacturing approaches are one part of a larger shift in how we deal with infectious pathogens from now on.

And there has been a substantial shift across the whole of vaccine and infectious disease research. I am still processing what I have learned from this pandemic and how it might shape my research going forward. One conclusion is that I know less now than I did at the beginning of 2020 – I am even more aware of gaps in my knowledge. I was preparing my lectures on vaccines for the 2021–22 academic year, and everything was out of date. All references to RNA and viral vectors as 'next-generation' vaccines had to go – they are the now generation! But the changes go deeper. One previous assumption was that, to be effective, a vaccine would need to provide 'sterilising immunity', i.e. act as a shield to stop virus from getting into the body. This in turn implied that the vaccine would need to induce a certain type of immune response, skewed towards antibodies that could 'kill'* the virus before it ever lived. Another assumption was that an effective vaccine could prevent the spread of a virus – through the (in)famous herd immunity. Whilst the COVID vaccines may do both of these things against the original strains, the coronavirus very sneakily evolved into the hyper-transmissible Omicron variant (so named for reasons best known to the WHO, skipping eleven Greek letters from Delta, including the pun-tastic Nu, pronounced new). The vaccines then provided less protection against infection. I can prove this with first-hand anecdotal evidence: I was infected with SARS-CoV-2 again in December 2021, less than eleven months after having had it the first time and despite three vaccine doses.

* Let's not go there again. See Chapter 2 (under Taxonomy).

And yet the vaccines have had an impressive impact on preventing disease – the peak of UK hospitalisations over the winter of 2021–22 was 15,000 per day, far lower than the predicted 45,000 a day. People who got infected after vaccination have tended to experience much milder symptoms, and it is worth remembering that without the vaccines they could have been much sicker. In my case, the first time I was infected I ended up bed-bound for a week, with a lasting impact on my ability to sing. The second infection presented as the mildest of colds, only discovered when I did a lateral flow test to go to work: most importantly, my ability to belt out 'God Rest Ye Merry Gentlemen' remained intact for the Christmas period. This 'what might have been' type of analysis is important. We need to think about vaccines differently, considering those that prevent disease alongside those that prevent infections and those that reduce transmission.

So what happens next? If I come back and revise the book in the future (and bear in mind my staggeringly low prediction success rate) what do I predict will have happened? We have learned an awful lot in the last two years. It has been an unprecedented opportunity to observe the course of a highly transmissible respiratory virus, with tools to interrogate every aspect: from how the virus behaves to how differences in individuals' genes shape their susceptibility to infection. The first fruits from this learning will come in the next generation of vaccines. There will be a huge acceleration in RNA vaccines research and development. The remarkable success of these vaccines against COVID opens up the possibility of their use against a range of diseases – though they may not be useful against all pathogens. We may also see the redistribution of vaccine manufacture: Cuba, for example, built on experience within its local biotech industry to develop a COVID vaccine called Abdala, which was 92% effective. This homegrown vaccine freed the country from competing with richer countries for the limited early production runs of vaccines. More generally, enhanced understanding about the way vaccination and the immune system protect against disease may improve vaccines against other respiratory viruses, opening the door for vaccines that can work against pathogens that mutate – in particular influenza. Reducing the burden of disease without repeat immunisations would both help reduce the impact of the seasonal winter viruses and ideally reduce the impact of any future pandemics.

Sadly, the predictions aren't all positive – COVID has sucked up resources from other research and interrupted implementation schemes of other vaccines: the eradication of polio has been pushed further back and the threat of antimicrobial resistant bacteria remains firmly on the horizon. But don't let these dark clouds on the horizon scare you too much. As with most of human progress, science is a mixed bag: for every two steps forward there will be one step back. But as the dance with disease continues, I remain optimistic that we will continue to learn more about pathogens and be able to use this knowledge to prevent the impact they have on us.

And so, I will end on one more reason to celebrate. Another pathogen looks like it will be eradicated: Guinea worm, a parasitic worm that can grow up to a metre long and is immensely painful as it exits the skin. In the 1980s, there were 3.5 million cases a year; in 2021, there were fourteen cases in total. If this happens, it will be the third pathogen to be eradicated (and the second human one). It is especially remarkable because there is no vaccine, so eradication is down to tracking infected individuals and preventing onward spread. Combining lower tech approaches with the next generation of vaccine technology paves the way for even more success. So while we are trapped in the pathogen two-step, we may eventually run out of dance partners, which is a cause to celebrate, especially if you have seen me dance!

Acknowledgements

There are so many people to thank for this, as it is a project fifteen years in the making: Alan Foster and Sarah Harman, who encouraged me into blogging; Jenny Rohn, who gave me a well-timed nudge to go global; Julie Gould and Chris Parr, for letting me write on websites read by more than just my family; Jack Leeming and Paul Jump, who gave me my first professional writing gigs; Dan Davis, who gave me the nudge and inspiration to write a book, with the helpful tip 'just don't make it shit'; Caroline Hardman, my amazing agent, for having the idea in the first place; all at Hardman & Swainson for guiding me through the weird world of publishing; and Sam Carter at Oneworld, for taking on my book (and educating me about comma splices), Dad Jokes and all.

The illustrations are thanks to the wonderfully talented Ash Uruchurtu and are supported by a Communicating Immunology grant from the British Society for Immunology. With thanks to my colleagues Michal S. Barski and Goedele N. Maertens for the protein crystals and diffusion plots and Sophie Higham for the bacterial Petri dish photo.

I am lucky to be surrounded by wonderful scientists whom I can tap up for ideas. With thanks to Steven Riley and Nick Grassly for filling in many of my epidemiology shortcomings; Andy Edwards for microbiology; Gordon Dougan for vaccines; Ian Humphreys for immunology; Mike Cox for microbiome tips; Blair Strang for virology; Kat Arney for dealing with conspiracies; Thushan da Silva for diagnostics; and Tom Sewell for fungal wisdom. I also need to thank Cecilia Johansson for tolerating me when I appear at her office door to chat about rubbish whenever I am bored. And a huge thanks goes to the hive mind of Twitter for answering random questions at odd times in the day. And, of course, my actual research would never happen without the amazing Team JT past and present – Simren, Ryan, Katya, Adam, Jaq, Miko, Dave, Helen, Meg, Vicky, Matt, Laura, Charanjit, Katie, Emily, Sophie, David, Zee, Jonny, Rhia and Felicity.

My personal scientific journey has been shaped by a huge cast of wonderful, inspirational mentors. Mr Parkin and Dr Cruse at school for opening the door of exploration; Stephen Russell for giving me my first insight of real research and laughing at my filthy jokes; Doog and Peter Nixon for supporting me through my PhD and ever since; Peter Openshaw for my first actual job and the continued insight; Robin Shattock for the last twelve years of friendship, guidance and support; and Wendy Barclay for being an inspirational boss.

I would not be writing this without my family. The book was in part inspired by the two historians in the family (my father Richard and my sister Clare) not really knowing any science but having a damned good try to make sense of the world (Pickled Onion gene?) and being better at telling the time. My mother Georgina and sister Emily have a slightly better grip on the world scientific (at least the maths side) and all four of them are the rock upon which this is built.

My final heartfelt thanks go to my toughest critic, my son Jamie, and my most adoring fan, my daughter Sophie. And none of these words would ever have been written without my wonderful wife, who every weekend for twelve months took care of the children when I locked myself in the study and ground out another two thousand words (all while she coordinated the COVID-19 vaccine response for the Wellcome Trust). Hopefully the occasional gift of gin and flowers made up in part for some of my absence.

Abbreviations

Abbreviation	
ADA	adenosine deaminase
AI	artificial intelligence
AIDS	acquired immune deficiency syndrome
AMR	anti-microbial resistance
AZT	azidothymidine (zidovudine): anti-HIV drug
BCG	Bacillus Calmette-Guérin
CCR5	C-C chemokine receptor 5
CD	cluster of differentiation
CDC	Centers for Disease Control (USA)
CF	cystic fibrosis
CFTR	cystic fibrosis transmembrane conductance regulator
CFU	colony forming unit
COPD	chronic obstructive pulmonary disease
COVID	coronaVIrus disease
CRISPR	clustered regularly interspaced short palindromic repeats
DNA	deoxyribonucleic acid
DRC	Democratic Republic of the Congo, formerly Zaire
EBV	Epstein-Barr virus
ELISpot	enzyme-linked immune absorbent spot
FDA	Food and Drug Agency (USA)
FIV	feline immunodeficiency virus
FMT	faecal microbiodata transplant
GAVI	Global Alliance for Vaccines and Immunization
GBS	group B Streptococcus
GSK	GlaxoSmithKline

HAART	highly active antiretroviral therapy
hCG	human chorionic gonadotropin
Hep	hepatitis (can be type A, B or C – HepA, HepB or HepC). Caused by hepatitis A virus (HAV), hepatitis B virus (HBV) or hepatitis C virus (HCV)
HiB	*Haemophilus influenzae* type B
HIV	human immunodeficiency virus
HLA	human leukocyte antigen
HPV	human papillomavirus
HSV	herpes simplex virus
HTLV	human T-lymphotropic virus
IDU	5-iodo-2′-deoxyuridine (an antiviral drug)
IFN	interferon
IFR	infection fatality rate
IL	interleukin
LB	lysogeny broth
LMIC	low- and middle-income countries
LPS	lipopolysaccharide
MALDI-TOF-MS	Matrix Assisted Laser Desorption Ionization-Time Of Flight Mass Spectrometry
MDR	multi-drug resistant bacteria
MHC	major histocompatibility complex
MMR	measles, mumps and rubella vaccine
MPL	monophosphoryl lipid
MRSA	methicillin resistant *Staphylococcus aureus*
MSM	men who have sex with men
ng	nanogram (1 billionth of a gram)
NHS	National Health Service (UK)
NIH	National Institutes of Health (USA)
NTD	neglected tropical diseases
PAMP	pathogen associated molecular pattern
PCR	polymerase chain reaction
PDR	pan-drug resistant bacteria
PhD	Doctor of Philosophy
PHE	Public Health England
PPE	personal protective equipment
PrEP	pre-exposure prophylaxis
RAG	recombination activating gene
RCT	randomised control trials
RNA	ribonucleic acid

R_0	basic reproduction number
RSV	respiratory syncytial virus
SAGE	Scientific Advisory Group for Emergencies
saRNA	self-amplifying RNA vaccine
SARS-CoV-2	severe acute respiratory syndrome-coronavirus-2
SCID	severe combined immunodeficiency
SIR	Susceptible-Infectious-Recovered
SIV	simian immunodeficiency virus
SPI-M	Scientific Pandemic Influenza Group on Modelling
STI	sexually transmitted infection
TB	tuberculosis
TCR	T cell receptor
TLA	Three letter acronym
TLR	toll-like receptor
TNF	tumour necrosis factor
μg	microgram (1 millionth of a gram)
UTI	urinary tract infection
WHO	World Health Organization
XDR	extensively drug resistant bacteria
XMRV	xenotropic murine leukaemia virus-related virus

Sort of Glossary

Anaemia: Tiredness and fatigue caused by the reduced ability of the blood to carry oxygen.

Antibody: Proteins produced by the immune system. They recognise the pathogens and bind to them very tightly. Basis for the protective effect of vaccines.

Antigen and epitope: Antigens are the bits of pathogens recognised by the immune system. The epitope is a smaller sub-region within the antigen that is specifically recognised.

B cells: White blood cells. Have a memory response – they recognise a specific antigen and release antibody in response.

Bacteria: Microorganism in the domain prokaryote. Has a cell wall and no nucleus. Incredibly diverse range of organisms living in all sorts of weird and wonderful places. Singular is technically bacterium, but there's no way I am going to use that.

Bacteriophage: Virus that infects bacteria.

Biofilm: Loads of bacteria living together in blissful harmony.

Cell lines: Cells normally have a limited lifespan and die. However, some cells like cancers can live forever. These cells can then be used to grow viruses.

Cell membrane: Lipid layer that encapsulates our cells (made in part of cholesterol – it's why you need some fat in your diet).

Cell wall: Thick outer layer surrounding a bacteria. Made of LPS and peptidoglycan amongst other things. Keeps the bacteria safely contained, but also triggers the immune system to danger.

Chemokine: A signalling protein. It is released from the site of infection to draw white blood cells to the pathogen.

Chirality: The mirror orientation of a molecule. Some molecules can be overlaid and others, because of the way they fold in a three-dimensional space, cannot be overlaid. This is similar to your left and right hands: they have mirror symmetry, but you cannot put one over the top of the other.

Commensals: Where two species interact and one gains and the other neither benefits nor loses. It is subtly different from mutualism, where both parties gain.

CRISPR-Cas: Next generation gene engineering tool, derived from bacteria; it can be targeted to any gene sequence.

Culture: Growth of a microorganism. Not dissimilar to the cultures found in 'live yoghurt'.

Cystic fibrosis (CF): This is caused by a mutation in a gene called CFTR. It leads to stickier mucus in the lungs and guts and predisposes individuals to more severe infections.

Cytokine: A protein that is released into the bloodstream, activating and instructing the function of white blood cells.

Efficacy: Protection against an infection by a drug or vaccine. It is normally presented as a percentage, indicating the number of people it works in.

Enteric: A posh word for 'of the gut', from the Greek for intestine. Other types of infection are grouped as GI (gastrointestinal: guts), GU (genito-urinary: willies) and URTI (upper-respiratory tract infections: lungs).

Enzymes: Biological molecules that can catalyse (accelerate) a chemical reaction. They enable reactions to happen at the body temperature. Different enzymes can catalyse different reactions, from breaking things down (often these enzymes have the suffix -ase) to building things up.

Eukaryotic: A broad class of organisms, characterised by membrane-bound organelles such as the nucleus. Humans are eukaryotes, as are trees, birds, yeast and microscopic worms.

Extremophiles: Bacteria that can live in harsh environments, e.g. *Thermophilus aquaticus* found at Mushroom Spring in the Lower Geyser Basin of Yellowstone National Park.

Fungi: Microscopic mushrooms. Mostly harmless unless your immune system is already a bit broken. Part of the same domain of life as humans, just much, much smaller.

Gene sequencing: It is possible to read the genetic code of an organism. A range of different approaches is now used. But they turn the chemistry of the DNA molecule into a string of letters.

HLA: Human leukocyte antigen; a cluster of genes that encodes the MHC. It determines your immunity to infections and is also important for transplant surgery. There are three loci for MHC-I – HLA-A, HLA-B and HLA-C – and three loci for MHC-II – HLA-DP, HLA-DQ and HLA-DR.

Hormone: Signalling molecules that control the body's physiology or behaviour; examples include insulin, adrenalin, testosterone and cortisone.

Interferon: The body's antiviral alarm system. Alerts surrounding cells and reduces the spread of virus in the body.

Lateral flow (antibody) tests: For example, pregnancy test. Uses antibody to detect a specific protein in the biological sample.

LMIC: The term LMIC is used in health contexts rather than developing country, which in turn replaced the term Third World. LMIC are defined by income. For the 2021 fiscal year, low-income economies were defined as those with a gross national income per capita, calculated using the World Bank Atlas method, of $1,035 or less in 2019; lower middle-income economies were those with a gross national income per capita between $1,036 and $4,045; upper middle-income economies were those with a gross national income per capita between $4,046 and $12,535; high-income economies were those with a gross national income per capita of $12,536 or more.

Meningitis: Infection of the brain, often striking in infancy, leading to lifelong paralysis.

MHC: Major histocompatibility complex. The protein product of HLA genes. They are found on the surface of cells and used as a system to display the contents of the cell to the immune system.

Microbiome: This is the sum total of all the bacteria living on or in a surface or tissue.

Monoclonal antibody: A single antibody that recognises a single target. Can be generated in bulk thanks to the technology developed by Köhler and Milstein.

Nucleic acid composition: DNA has four constituent bases, represented by the letters A (adenosine), T (thymidine), C (cytosine) and G (guanosine). DNA is a double helix, with two molecules intertwined: A pairs with T and C with G. This means that identical copies can be made of the molecule.

Nucleotide/Nucleoside: Building block of DNA/RNA. There is a subtle difference (nucleosides lack the phosphate group) that will bother biochemists and virologists, but I use nucleotide throughout.

Nucleus: The instruction centre of eukaryotic cells, which contains DNA wrapped around a protein called chromatin.

PAMP: Pathogen associated molecular pattern. Biochemicals found in pathogens but not in human cells. They trigger the immune system to react. One example is LPS, which makes up the bacterial cell wall.

Parasites: In general, one organism that lives off another; in infection, very diverse family of eukaryotic micro-organisms that infect tissues.

Passive immunisation: Preventing infection by transferring antibodies from one person who has previously been infected to a susceptible / at risk person.

Pathogen: A microorganism that causes infection.

PCR (Polymerase Chain Reaction): A method of artificially amplifying a specific piece of DNA. It can be used to detect very tiny amounts of genetic material – for example in a COVID-infected nasal swab.

Plasmids: Mobile DNA element that bacteria have in addition to the main source of genetic information in their chromosomes, which is often presented as circles in textbooks. Useful for genetic engineering.

Prodrug: A drug that needs the body to alter it somehow (metabolise it) to become activated. This can reduce toxicity or improve absorption.

Prospective: A clinical study where you choose a population and look for the incidence of disease over time.

Proteins: These are the workhorses of life. They make the structures that build our bodies, the enzymes that catalyse reactions and the signals that let cells talk to each other. The instructions for each protein are encoded in the DNA.

Proteolytic: Enzymes that chew up proteins.

Randomised control trial: A clinical experiment with two or more groups, where you can compare the effect of your intervention. It is normally blinded.

Receptor: A protein found on the surface of a cell that sends a signal into the cell when it is activated by another molecule (called a ligand). Basically, a swipe-card reader on the cell that is only activated by specific protein.

Recombinant technology: Cloning the genes from one organism to make a protein in another. Used to make vaccines and human insulin.

Red Queen hypothesis: How two competing organisms can co-evolve but never really gain an advantage. It comes from *Through the Looking-Glass*, where Alice and the Red Queen ran incredibly fast but didn't get anywhere.

Restriction enzyme: Enzyme used by bacteria to cut specific site within DNA as a form of defence. It is used by scientists to engineer genes.

Retrospective: A clinical study where you look for the incidence of disease over time and then look to see whom it has affected.

Ribosomes: Cellular protein factories. They work a bit like 3D printers, translating RNA into protein.

Rice-water: A highly liquid stool, characteristic of cholera. Named because it looks a bit like water that rice has been cooked in (greyish/watery).

R_0: Replication number of an infection that measures how rapidly a pathogen spreads in a population. It is the number of people that one person infects.

Science Media Centre: A truly heroic organisation, the SMC was set up following the GM food debate to enable the interface between scientists and the society they serve. It pairs up journalists with experts.

Sensitivity and specificity: The rate at which a test returns a true positive result is sometimes referred to as the sensitivity of the test – it is the number of true positives divided by the total number of positive results. The rate at which a test returns a negative result is referred to as the specificity of the test – it is the number of true negatives divided by the total number of negative results.

Sepsis: Out of control immune reaction to infection leading to multi-organ failure. Most commonly triggered by a bacterial infection, particularly if it gets into the blood.

Serotype: A strain of an individual pathogen species that can be differentiated by whether a specific antibody binds to it.

SI Units: These are the standard units used by all scientists. They are probably unpopular with Brexiteers and other people stuck in the nineteenth century. They are called SI from the French – *Système Internationale d'unités* – after the first international conference on weights and measures, the Metre Convention in Paris. They include the gram, the second, the metre and the kelvin. They are easier to use and more universal than imperial measurements, but there have been notable cock-ups, e.g. the Mars orbiter. We then add a system of prefixes to say whether they are larger or smaller:

Name	Symbol	Base 10	Decimal	Word
giga	G	10^9	1000000000	billion
mega	M	10^6	1000000	million
kilo	k	10^3	1000	thousand
hecto	h	10^2	100	hundred
deca	da	10^1	10	ten
		10^0	1	one
deci	d	10^{-1}	0.1	tenth
centi	c	10^{-2}	0.01	hundredth
milli	m	10^{-3}	0.001	thousandth
micro	μ	10^{-6}	0.000001	millionth
nano	n	10^{-9}	0.000000001	billionth

Note that we sometimes present numbers as 10 with a superscript number following (e.g. 10^3) – this is called 10 to the power. The number in superscript is the number of zeros before or after the 1, so 10^3 would be 1,000.

Stem cell: Most cells are fixed in their fate; a skin cell cannot become a nerve cell, but stem cells have the potential to turn into any other type of cells.

Stochastic: A posh word for chance.

T cells: White blood cells. They have a memory response – they recognise a specific antigen and either kill the infected cell (CD8 T cells) or orchestrate the immune response to the pathogen (CD4 T cells).

Taxonomy: How things are grouped. For living things we use the Linnaean taxonomy, which goes Kingdom, Phylum, Class, Order, Family, Genus, Species (if you struggle to remember it, try this: King Phillip Came Over From Great Spain).

Transcription: How the genetic information gets from the DNA in the nucleus to the protein-making machinery of the cell. The information is transcribed from the DNA into RNA molecules. Because of the way the pairs form, this means that the RNA can mirror what is written in the DNA.

Translation: How the information in RNA is turned into proteins. The RNA goes to the protein machinery of the cells, called ribosomes. These read the sequence and add amino acids one at a time to the protein, a bit like Lego blocks. Each set of three letters (triplets) in the RNA instructs the ribosome to add a specific amino acid.

Virus: Obligate parasite. Must infect cells to make copies of itself. Very simple form of microorganism. Probably not alive at all, but don't tell the virologists that.

Zoonosis: An infection that spreads from other animal species to humans.

Notes

PROLOGUE: ENTER A VIRUS STAGE-EAST

1 The Great Death, in *The Times of London*, 1921.
2 *Office for National Statistics: Deaths registered weekly in England and Wales, provisional* 2020; Available from: https://www.ons.gov.uk/peoplepopulationandcommunity/birthsdeathsandmarriages/deaths/datasets/weeklyprovisionalfiguresondeathsregisteredinenglandandwales
3 WHO, *Global health estimates: Leading causes of death* 2019; Available from: https://www.who.int/data/gho/data/themes/mortality-and-global-health-estimates/ghe-leading-causes-of-death
4 Zhu, N., *et al.*, A Novel Coronavirus from Patients with Pneumonia in China, 2019. New England Journal of Medicine, 2020. **382**(8): pp. 727–33.
5 Wheeler, A. and W.R. Jack, *Wheeler and Jack's handbook of medicine*. 11th ed. 1932, Baltimore: Williams and Wilkins.
6 Rosling, H., O. Rosling and A.R. Rönnlund, *Factfulness: Ten reasons we're wrong about the world – and why things are better than you think*. First ed. 2018, New York: Flatiron Books.

CHAPTER 1: A BRIEF HISTORY OF MICROBIOLOGY

1 Dawkins, R., *The selfish gene*. New ed. 1989, Oxford; New York: Oxford University Press.
2 Diamond, J.M., *Guns, germs, and steel: the fates of human societies*. 1997, New York: W.W. Norton.
3 Peng, Y., *et al.*, The ADH1B Arg47His polymorphism in east Asian populations and expansion of rice domestication in history. BMC evolutionary biology, 2010. **10**: DOI: 10.1186/1471-2148-10-15

CHAPTER 2: THE SCIENCE OF MICROBIOLOGY

1 Bossert, B. and K.K. Conzelmann, Respiratory syncytial virus (RSV) nonstructural (NS) proteins as host range determinants: a chimeric bovine RSV with NS genes from human RSV is attenuated in interferon-competent bovine cells. J Virol, 2002. **76**(9): pp. 4287–93.
2 Whitfield, J., Portrait of a Serial Killer. Nature, 3 October 2002.

CHAPTER 3: WHY DON'T WE GET SICK?

1 Siggins, M.K., *et al.*, PHiD-CV induces anti-Protein D antibodies but does not augment pulmonary clearance of nontypeable *Haemophilus influenzae* in mice. Vaccine, 2015. **33**(38): pp. 4954–61.
2 Fowke, K.R., *et al.*, Resistance to HIV-1 infection among persistently seronegative prostitutes in Nairobi, Kenya. The Lancet, 1996. **348**(9038): pp. 1347–51.
3 Weinberg, E.D., Nutritional Immunity: Host's Attempt to Withhold Iron From Microbial Invaders. JAMA, 1975. **231**(1): pp. 39–41.
4 Sazawal, S., *et al.*, Effects of routine prophylactic supplementation with iron and folic acid on admission to hospital and mortality in preschool children in a high malaria transmission setting: community-based, randomised, placebo-controlled trial. The Lancet, 2006. **367**(9505): pp. 133–43.
5 Gill, S.K., *et al.*, Increased airway glucose increases airway bacterial load in hyperglycaemia. Sci Rep, 2016. **6**: DOI: 10.1038/srep27636

CHAPTER 4: THE MICROBIOME

1 Sender, R., S. Fuchs and R. Milo, Revised Estimates for the Number of Human and Bacteria Cells in the Body. PLoS biology, 2016. **14**(8): DOI: 10.1371/journal.pbio.1002533
2 Falony, G., *et al.*, Population-level analysis of gut microbiome variation. Science, 2016. **352**(6285): pp. 560–4.
3 Thaiss, C., *et al.*, Transkingdom Control of Microbiota Diurnal Oscillations Promotes Metabolic Homeostasis. Cell, 2014. **159**(3): pp. 514–29.
4 Groves, H.T., *et al.*, Respiratory Viral Infection Alters the Gut Microbiota by Inducing Inappetence. mBio, 2020. **11**(1): DOI: 10.1128/mBio.03236-19

CHAPTER 5: IMMUNOLOGY

1 Busse, D.C., *et al.*, Interferon-Induced Protein 44 and Interferon-Induced Protein 44-Like Restrict Replication of Respiratory Syncytial Virus. J Virol, 2020. **94**(18).

2 Janeway, C., *Immunobiology: the immune system in health and disease*. 5th ed. 2005, New York: Garland Science.

3 Matzinger, P. and G. Mirkwood, In a fully H-2 incompatible chimera, T cells of donor origin can respond to minor histocompatibility antigens in association with either donor or host H-2 type. J Exp Med, 1978. **148**(1): pp. 84–92.

4 Chiou, K.L. and C.M. Bergey, Methylation-based enrichment facilitates low-cost, noninvasive genomic scale sequencing of populations from feces. Sci Rep, 2018. **8**(1): p. 1975.

5 Crichton, M., *The Andromeda strain*. 1969, New York: Knopf.

6 Zhang, Q., *et al.*, Inborn errors of type I IFN immunity in patients with life-threatening COVID-19. Science, 2020. **370**(6515): DOI: 10.1126/science.abd4570

7 Davis, D.M., *The Compatibility Gene: How Our Bodies Fight Disease, Attract Others, and Define Our Selves*. 2014, Oxford: Oxford University Press.

8 Miura, T., *et al.*, HLA-B57/B*5801 Human Immunodeficiency Virus Type 1 Elite Controllers Select for Rare Gag Variants Associated with Reduced Viral Replication Capacity and Strong Cytotoxic T-Lymphotye Recognition. J Virol, 2009. **83**(6): pp. 2743–55.

9 Dick, P.K., *Do androids dream of electric sheep?* 1968, Garden City, NY: Doubleday.

10 Briney, B., *et al.*, Commonality despite exceptional diversity in the baseline human antibody repertoire. Nature, 2019. **566**(7744): pp. 393–7.

11 Cole, M.E., *et al.*, Pre-existing influenza specific nasal IgA or nasal viral infection does not affect live attenuated influenza vaccine immunogenicity in children. Clin Exp Immunol, 2020. DOI: 10.1111/cei.13564

12 Gould, V.M.W., *et al.*, Nasal IgA Provides Protection against Human Influenza Challenge in Volunteers with Low Serum Influenza Antibody Titre. Front Microbiol, 2017. **8**: p. 900.

CHAPTER 6: EPIDEMIOLOGY

1 du Plessis, L., *et al.*, Establishment and lineage dynamics of the SARS-CoV-2 epidemic in the UK. medRxiv, 2020: DOI: 10.1101/2020.10.23.20218446

2 Kahneman, D., *Thinking, fast and slow*. 2011, New York: Farrar, Straus and Giroux.

3 Bedingfield, W., Was Cheltenham a coronavirus super-spreader event?, in *Wired*, 2020.

CHAPTER 7: DIAGNOSTICS

1 Lepore, J., It's Spreading, in *The New Yorker*, 2009.

2 Maniatis, T., E.F. Fritsch and J. Sambrook, *Molecular cloning: a laboratory manual*. 1982, Cold Spring Harbor, NY: Cold Spring Harbor Laboratory.

3 Skloot, R., *The immortal life of Henrietta Lacks*. 2010, New York: Crown Publishers.

4 Johnson, G., Bright Scientists, Dim Notions, in *The New York Times*, 2007.

5 Gabbatiss, J., *James Watson: The most controversial statements made by the father of DNA*. 2019; Available from: https://www.independent.co.uk/news/science/james-watson-racism-sexism-dna-race-intelligence-genetics-double-helix-a8725556.html

6 van Kuppeveld, F.J.M. and J.W.M. van der Meer, XMRV and CFS – the sad end of a story. The Lancet, 2012. 379(9814): pp. e27–e28.

7 Cohen, J., *Dispute Over Lab Notebooks Lands Researcher in Jail*. Science 2011; Available from: https://science.sciencemag.org/content/334/6060/1189/

8 Chavarria-Miró, G., *et al.*, Sentinel surveillance of SARS-CoV-2 in wastewater anticipates the occurrence of COVID-19 cases. medRxiv, 2020: DOI: 10.1101/2020.06.13.20129627

9 *COG-UK passes 100K genomes*. 2020; Available from: https://www.cogconsortium.uk/news_item/cog-uk-passes-100k-genomes/

10 Cohen, A., *et al.*, The global prevalence of latent tuberculosis: a systematic review and meta-analysis. European Respiratory Journal, 2019: DOI: 10.1183/13993003.00655-2019

CHAPTER 8: PREVENTION

1 Wheelis, M., Biological Warfare at the 1346 Siege of Caffa. Emerging Infectious Diseases, 2002. **8**(9): p. 971.

2 Kwok, Y.L., J. Gralton and M.L. McLaws, Face touching: a frequent habit that has implications for hand hygiene. Am J Infect Control, 2015. **43**(2): pp. 112–14.

3 Atchison, C.J., *et al.*, Perceptions and behavioural responses of the general public during the COVID-19 pandemic: A cross-sectional survey of UK Adults. medRxiv, 2020: DOI: 10.1101/2020.04.01.20050039

4 Miranda, R.C. and D.W. Schaffner, Longer Contact Times Increase Cross-Contamination of *Enterobacter aerogenes* from Surfaces to Food. Applied and Environmental Microbiology, 2016. **82**(21): pp. 6490–6.

5 Hirose, R., *et al.*, Situations Leading to Reduced Effectiveness of Current Hand Hygiene against Infectious Mucus from Influenza Virus-Infected Patients. mSphere, 2019. **4**(5): DOI: 10.1128/mSphere.00474-19

6 Iversen, B., *et al.*, Should individuals in the community without respiratory symptoms wear facemasks to reduce the spread of COVID-19? 2020: Norwegian Institute of Public Health.

7 Bundgaard, H., *et al.*, Effectiveness of Adding a Mask Recommendation to Other Public Health Measures to Prevent SARS-CoV-2 Infection in Danish Mask Wearers. Annals of Internal Medicine, 2020.

8 WHO, *El Salvador certified as malaria-free by WHO*. 2021; Available from: https://www.who.int/news/item/25-02-2021-el-salvador-certified-as-malaria-free-by-who

9 CDC, Achievements in Public Health, 1900–1999: Control of Infectious Diseases. MMWR, 1999. **48**(29): pp. 621–9.

10 Osseiran, N., *1 in 3 people globally do not have access to safe drinking water – UNICEF, WHO*. 2019; Available from: https://www.who.int/news/item/18-06-2019-1-in-3-people-globally-do-not-have-access-to-safe-drinking-water-unicef-who

11 Masters, R., *et al.*, Return on investment of public health interventions: a systematic review. Journal of Epidemiology and Community Health, 2017. **71**(8): p. 827.

CHAPTER 9: VACCINES

1 Boylston, A.W., The Myth of the Milkmaid. New England Journal of Medicine, 2018. **378**(5): pp. 414–15.

2 Shane, S., Portrait Emerges of Anthrax Suspect's Troubled Life, in *The New York Times*, 2009.

3 Smith, K.A., Wanted, an Anthrax vaccine: Dead or Alive? Medical Immunology, 2005. **4**(1): DOI: 10.1186/1476-9433-4-5

4 Wright, A.E., *The unexpurgated case against woman suffrage*. 1913, New York: P.B. Hoeber.

5 Oakley, C.L., Alexander Thomas Glenny, 1882–1965. *Biographical memoirs of fellows of the Royal Society*, 1966.

6 Breman, J.G., *et al.*, Discovery and Description of Ebola Zaire Virus in 1976 and Relevance to the West African Epidemic During 2013–2016. The Journal of Infectious Diseases, 2016. **214**(suppl_3): pp. S93–S101.

7 Tregoning, J.S., *et al.*, Vaccines for COVID-19. Clin Exp Immuno, 2020. **202**(2): pp. 162–92.

8 Polack, F.P., *et al.*, Safety and Efficacy of the BNT162b2 mRNA Covid-19 Vaccine. New England Journal of Medicine, 2020. **383**(27): pp. 2603–15.

9 Vogel, A.B., *et al.*, Self-Amplifying RNA Vaccines Give Equivalent Protection against Influenza to mRNA Vaccines but at Much Lower Doses. Mol Ther, 2018. **26**(2): pp. 446–55.

10 Cohen, J., Studies that intentionally infect people with disease-causing bugs are on the rise, in Science, 2016.

11 Almeida, J.D. and D.A.J. Tyrrell, The Morphology of Three Previously Uncharacterized Human Respiratory Viruses that Grow in Organ Culture. Journal of General Virology, 1967. **1**(2): pp. 175–8.

12 Kramer, J., They spent 12 years solving a puzzle. It yielded the first COVID-19 vaccines, in *National Geographic*, 2020.

13 Zhong, Z., *et al.*, The impact of timing of maternal influenza immunization on infant antibody levels at birth. Clin Exp Immunol, 2019. **195**(2): pp. 139–52.

14 Ramon, G., Sur la toxine et sur l'anatoxine diphtheriques. Ann. Inst. Pasteur, 1924. **38**(1).

15 Russell, R.F., *et al.*, Use of the Microparticle Nanoscale Silicon Dioxide as an Adjuvant To Boost Vaccine Immune Responses against Influenza Virus in Neonatal Mice. J Virol, 2016. **90**(9): pp. 4735–44.

16 Palmer, T., *et al.*, Prevalence of cervical disease at age 20 after

immunisation with bivalent HPV vaccine at age 12–13 in Scotland: retrospective population study. BMJ, 2019. **365**: p. l1161.

17 CDC, Nonfatal Bathroom Injuries Among Persons Aged ≥15 Years – United States, 2008. Morbidity and Mortality Weekly Report, 2008. **60**(22): pp. 729–33.

18 WHO. *First ever vaccine listed under WHO emergency use*. 2020; Available from: https://www.who.int/news/item/13-11-2020-first-ever-vaccine-listed-under-who-emergency-use

19 Parker, A.A., *et al.*, Implications of a 2005 Measles Outbreak in Indiana for Sustained Elimination of Measles in the United States. New England Journal of Medicine, 2006. **355**(5): pp. 447–55.

20 de Figueiredo, A., *et al.*, Mapping global trends in vaccine confidence and investigating barriers to vaccine uptake: a large-scale retrospective temporal modelling study. The Lancet, 2020. **396**(10255): pp. 898–908.

21 McDonald, J.U., *et al.*, Inflammatory responses to influenza vaccination at the extremes of age. Immunology, 2017. **151**(4): pp. 451–63.

22 Thaler, R.H. and C.R. Sunstein, *Nudge: Improving decisions about health, wealth, and happiness*. 2008, New Haven: Yale University Press.

23 Patel, M., Test behavioural nudges to boost COVID-19 immunization. Nature, 2021. **590**(7845): p. 185.

24 Trust, T.W., Wellcome Global Monitor. 2018.

25 Hviid, A., *et al.*, Association between thimerosal-containing vaccine and autism. JAMA, 2003. **290**(13): pp. 1763–6.

CHAPTER 10: ANTIBIOTIC

1 Summers, W.C., The strange history of phage therapy. Bacteriophage, 2012. **2**(2): pp. 130–3.

2 Smith, M.L., Honey bee sting pain index by body location. PeerJ, 2014. **2**: p. e338.

3 Unger, D.L., Does knuckle cracking lead to arthritis of the fingers? Arthritis Rheum, 1998. **41**(5): pp. 949–50.

4 Pathak, A., *et al.*, Comparative genomics of Alexander Fleming's original Penicillium isolate (IMI 15378) reveals sequence divergence of penicillin synthesis genes. Sci Rep, 2020. **10**(1): DOI: 10.1038/s41598-020-72584-5

5 Steinmetz, K., Esther Lederberg and Her Husband Were Both Trailblazing Scientists. Why Have More People Heard of Him?, in *Time*, 2019.

6 Landers, T.F., *et al.*, A review of antibiotic use in food animals:

perspective, policy, and potential. Public health reports (Washington, DC: 1974), 2012. **127**(1): pp. 4–22.

7 Ling, L.L., *et al.*, A new antibiotic kills pathogens without detectable resistance. Nature, 2015. **517**(7535): pp. 455–9.

CHAPTER 11: ANTIVIRAL

1 Gallo, R., *et al.*, Frequent detection and isolation of cytopathic retroviruses (HTLV-III) from patients with AIDS and at risk for AIDS. Science, 1984. **224**(4648): pp. 500–3.

2 Barré-Sinoussi, F., *et al.*, Isolation of a T-lymphotropic retrovirus from a patient at risk for acquired immune deficiency syndrome (AIDS). Science, 1983. **220**(4599): pp. 868–71.

3 Faria, N.R., *et al.*, The early spread and epidemic ignition of HIV-1 in human populations. Science, 2014. **346**(6205): p. 56.

4 Pépin, J., *The origins of AIDS*. 2011, Cambridge, UK; New York: Cambridge University Press.

5 Fischl, M.A., *et al.*, The Efficacy of Azidothymidine (AZT) in the Treatment of Patients with AIDS and AIDS-Related Complex. New England Journal of Medicine, 1987. **317**(4): pp. 185–91.

6 Hayes, R.J., *et al.*, Effect of Universal Testing and Treatment on HIV Incidence — HPTN 071 (PopART). New England Journal of Medicine, 2019. **381**(3): pp. 207–18.

7 Brown, T.R., I am the Berlin patient: a personal reflection. AIDS research and human retroviruses, 2015. **31**(1): pp. 2–3.

8 Gupta, R.K., *et al.*, Evidence for HIV-1 cure after CCR5Δ32/Δ32 allogeneic haemopoietic stem-cell transplantation 30 months post analytical treatment interruption: a case report. The Lancet HIV, 2020. **7**(5): pp. e340–e347.

9 Ledford, H., The unsung heroes of the Nobel-winning hepatitis C discovery, in Nature, 2020. **586**(7830): p. 485.

10 Beigel, J.H., *et al.*, Remdesivir for the Treatment of Covid-19 – Final Report. New England Journal of Medicine, 2020. **383**(19): pp. 1813–26.

11 Dyer, O., Covid-19: Remdesivir has little or no impact on survival, WHO trial shows. BMJ, 2020. **371**: p. m4057.

12 Jefferson, T., *et al.*, Neuraminidase inhibitors for preventing and treating influenza in healthy adults and children. The Cochrane database of systematic reviews, 2014. **2014**(4): DOI: 10.1002/14651858.CD008965.pub4

CHAPTER 12: ANTIFUNGAL

1 Xie, H.Y., *et al.*, Probiotics for vulvovaginal candidiasis in non-pregnant women. The Cochrane database of systematic reviews, 2017. **11**(11): DOI: 10.1002/14651858.CD010496.pub2

CHAPTER 13: ANTIPARASITIC

1 Poinar, G., Jr., *Plasmodium dominicana* n. sp. (*Plasmodiidae: Haemospororida*) from Tertiary Dominican amber. Syst Parasitol, 2005. **61**(1): pp. 47–52.
2 Welsh, I., *Trainspotting*. 1994, London: Minerva.
3 Weller, C.L., Anti-inflammatory proteins secreted by parasitic nematodes, in *Biochemistry, Imperial College*. 2002, University of London.
4 Richens, J., Genital manifestations of tropical diseases. Sexually Transmitted Infections, 2004. **80**(1): p. 12.
5 Janse, J.J., *et al.*, Establishing the Production of Male Schistosoma mansoni Cercariae for a Controlled Human Infection Model. The Journal of Infectious Diseases, 2018. **218**(7): pp. 1142–6.

CHAPTER 14: BAD MEDICINE

1 Mandel, E.S.J., *Station eleven*. 2015, New York: Alfred A. Knopf.
2 Chokshi, N., That Wasn't Mark Twain: How a Misquotation Is Born, in *The New York Times*, 2017.
3 Cook, J., *et al.*, *Coronavirus, 'Plandemic' and the seven traits of conspiratorial thinking*. November 2020; Available from: https://theconversation.com/coronavirus-plandemic-and-the-seven-traits-of-conspiratorial-thinking-138483
4 Lorch, M. *Why people believe in conspiracy theories – and how to change their minds*. 2017; Available from: https://theconversation.com/why-people-believe-in-conspiracy-theories-and-how-to-change-their-minds-82514
5 Douglas, K.M., *et al.*, Understanding Conspiracy Theories. Political Psychology, 2019. **40**(S1): pp. 3–35.
6 Sagan, C., *The Demon-Haunted World: Science as a Candle in the Dark*. 1995, New York: Random House.
7 Group, R.T., Effect of Hydroxychloroquine in Hospitalized Patients with Covid-19. New England Journal of Medicine, 2020. **383**(21): pp. 2030–40.

8 *Sorry sunshine, wrong place*. 2020; Available from: http://reuters.com/video/watch/idOVD08UUYP

CHAPTER 15: THE FUTURE

1 Kaforou, M., *et al.*, Diagnosis of Bacterial Infection Using a 2-Transcript Host RNA Signature in Febrile Infants 60 Days or Younger. JAMA, 2017. **317**(15): pp. 1577–8.
2 Natarajan, A., H.-W. Su and C. Heneghan, Assessment of physiological signs associated with COVID-19 measured using wearable devices. medRxiv, 2020: DOI: 10.1101/2020.08.14.20175265
3 Ahmad, I., R. Flanagan and K. Staller, Increased Internet Search Interest for GI Symptoms May Predict COVID-19 Cases in US Hotspots. Clinical Gastroenterology and Hepatology, 2020. **18**(12): pp. 2833–4.e3.
4 Carreyrou, J., *Bad blood: secrets and lies in a Silicon Valley startup*. 2018, New York: Alfred A. Knopf.
5 Booth, C., *et al.*, Gene therapy for primary immunodeficiency. Human Molecular Genetics, 2019. **28**(R1): pp. R15–R23.
6 Long, K.C., *et al.*, Core commitments for field trials of gene drive organisms. Science, 2020. **370**(6523): p. 1417.
7 Davis, D.M., *The beautiful cure: the revolution in immunology and what it means for your health*. 2018, Chicago: The University of Chicago Press.
8 Bullmore, E.T., *The inflamed mind: a radical new approach to depression*. 2019, New York: Picador.
9 Taquet, M., *et al.*, Bidirectional associations between COVID-19 and psychiatric disorder: retrospective cohort studies of 62,354 COVID-19 cases in the USA. The Lancet Psychiatry, 2021. 8(2): 130–40.
10 Tregoning, J.S., *et al.*, Expression of tetanus toxin Fragment C in tobacco chloroplasts. Nucleic Acids Res, 2003. **31**(4): pp. 1174–9.
11 Ward, B.J., *et al.*, Efficacy, immunogenicity, and safety of a plant-derived, quadrivalent, virus-like particle influenza vaccine in adults (18–64 years) and older adults (≥65 years): two multicentre, randomised phase 3 trials. The Lancet, 2020. **396**(10261): pp. 1491–1503.

EPILOGUE: WHERE DO WE GO FROM HERE?

1 Taleb, N.N., *The black swan: the impact of the highly improbable*. 2007, New York: Random House.

Index

References to images are in *italics*.

© Andy Pritchard

DR JOHN S. TREGONING is a scientist and researcher whose work focuses on how viruses and bacteria infect the lungs and how our immune system fights them. He is currently reader in respiratory infections at Imperial College London, where he runs a research group on infectious diseases. John has published over sixty academic papers and has written articles for *Times Higher Education* and *Science*. During the COVID-19 pandemic, he had a regular column in *Nature*. He has been interviewed by outlets including the BBC and the *Sunday Telegraph*, and appeared in the 2021 Royal Institution Christmas Lectures. @DrTregoning